AF798126

Palgrave Studies in Impact Finance

Series Editor
Mario La Torre, Facoltà di Economia, Department of Management,
Sapienza University of Rome, Rome, Italy

The *Palgrave Studies in Impact Finance* series provides a valuable scientific 'hub' for researchers, professionals and policy makers involved in Impact finance and related topics. It includes studies in the social, political, environmental and ethical impact of finance, exploring all aspects of impact finance and socially responsible investment, including policy issues, financial instruments, markets and clients, standards, regulations and financial management, with a particular focus on impact investments and microfinance.

Titles feature the most recent empirical analysis with a theoretical approach, including up to date and innovative studies that cover issues which impact finance and society globally.

More information about this series at
http://www.palgrave.com/gp/series/14621

Magdalena Ziolo · Beata Zofia Filipiak ·
Blanka Tundys

Sustainability in Bank and Corporate Business Models

The Link between ESG Risk Assessment and Corporate Sustainability

palgrave
macmillan

Magdalena Ziolo
Institute of Economics and Finance
University of Szczecin
Szczecin, Poland

Beata Zofia Filipiak
Institute of Economics and Finance
University of Szczecin
Szczecin, Poland

Blanka Tundys
Institute of Management
University of Szczecin
Szczecin, Poland

ISSN 2662-5105 ISSN 2662-5113 (electronic)
Palgrave Studies in Impact Finance
ISBN 978-3-030-72097-1 ISBN 978-3-030-72098-8 (eBook)
https://doi.org/10.1007/978-3-030-72098-8

© The Editor(s) (if applicable) and The Author(s), under exclusive license to Springer Nature Switzerland AG 2021
This work is subject to copyright. All rights are solely and exclusively licensed by the Publisher, whether the whole or part of the material is concerned, specifically the rights of translation, reprinting, reuse of illustrations, recitation, broadcasting, reproduction on microfilms or in any other physical way, and transmission or information storage and retrieval, electronic adaptation, computer software, or by similar or dissimilar methodology now known or hereafter developed.
The use of general descriptive names, registered names, trademarks, service marks, etc. in this publication does not imply, even in the absence of a specific statement, that such names are exempt from the relevant protective laws and regulations and therefore free for general use.
The publisher, the authors and the editors are safe to assume that the advice and information in this book are believed to be true and accurate at the date of publication. Neither the publisher nor the authors or the editors give a warranty, expressed or implied, with respect to the material contained herein or for any errors or omissions that may have been made. The publisher remains neutral with regard to jurisdictional claims in published maps and institutional affiliations.

Cover illustration: Westend61 GmbH/Alamy Stock Photo

This Palgrave Macmillan imprint is published by the registered company Springer Nature Switzerland AG
The registered company address is: Gewerbestrasse 11, 6330 Cham, Switzerland

Acknowledgments

Research results presented in this volume are an element of research project financed by the National Science Center Poland (NCN) OPUS16 2018/31/B/HS4/00570. The authors would like to thank the reviewer Souhaila Kammoun for her comments that helped to improve the monograph.

Contents

List of Figures

List of Tables

CHAPTER 1

Introduction

ESG risk and sustainability are gaining increased importance, especially in light of COVID-19. Finance continues to be a driver for sustainability, and recent regulations have forced banks and companies to transform their business models toward sustainability. The sustainable business models of banks and companies have so far been discussed separately, but this book looks to combine them by answering the following questions: Do banks impact corporate sustainability? How do banks impact company business models? How do banks and companies create sustainable value? How do banks impact company ESG risk?

The activities of banks are increasingly affected by non-financial factors: the environmental, social, and governance (ESG) factors. The risk accompanying ESG factors requires adjusting the activities of banks related to risk management to new conditions. ESG risk is also incurred by entrepreneurs and clients of banks, who transfer this risk to banks. Thus, dependencies and interactions in the field of ESG risk arise between banks and enterprises. Banks respecting the ESG criteria may take various actions affecting entrepreneurs, including completely excluding certain sectors from cooperation (e.g., representatives of dirty business), differentiating the cost of the service (most often, this is the cost of obtaining financing), demanding additional security, and/or increasing the frequency and scope of transaction monitoring. As a result of these activities, entrepreneurs may decide to introduce adaptive changes

© The Author(s), under exclusive license to Springer Nature Switzerland AG 2021

M. Ziolo et al., *Sustainability in Bank and Corporate Business Models*, Palgrave Studies in Impact Finance,
https://doi.org/10.1007/978-3-030-72098-8_1

allowing for more accurate adjustment of the business model to ESG criteria and thus increase the availability of financial services. A gap is created by these issues in the links between the strategies and business models of banks and enterprises in the field of sustainability. In particular, the gap involves the effectiveness of new business models and instruments and the creation of new value (the green value). It is often necessary to look for answers regarding the extent to which business models and the offer of banking products affect the behavior and shape of business models of enterprises. Do banks really influence the behavior of their clients through green instruments? Are banks' business strategies that consider sustainability able to effectively change their clients' approaches? The corporate sector is increasingly reconfiguring its business strategies toward a more sustainable approach. The actions taken relate to the context of doing business; the perception of the organization in the market; its outputs, including products; and to the context of adapting to national or international requirements and regulations. The financial sector plays an important role in the transformation process, which supports the activities of the sector. Guidelines and global trends indicate that the financial sector is increasingly inclined to support sustainable activities by offering such financial products. The EU taxonomy is both a supportive and guiding system, containing a catalog of technical criteria to be met by organizations. These guidelines are not only important for the organizations, but are also a starting point for the financial sector to support and finance their activities. The guidelines of the EU taxonomy, and the new and sustainable business models, courses of action, and created financial instruments for the corporate sector have a common goal: a more sustainable and less-polluted economy. The goals of environmental neutrality can only be achieved by acting together, and the indicated documents, regulations, instruments, and strategies serve to fulfill these goals. This book covers the research gap of the link between sustainable value creation and the sustainable business models of companies and banks. A summary precedes each chapter in the book and conclusions end each chapter. The aim of the book is to provide an assessment of the impact of banks on sustainable value creation in companies' business models. The book consists of eight chapters. This Chapter is introduction. Chapter 2 shows the sustainability impact of the concept of business models. In this section, we show why the sustainability approach is important for building business models. The chapter presents how banks should approach the construction of their business models, considering their internal specificity, creating original values, the factor of social responsibility, and ESG risk factors. The end result of this chapter is to answer to the question regarding the differences in designing business model canvases and how to approach sustainable business models.

Chapter 3 addresses the how banking regulations and restrictions on risk affect the ability to better integrate banking ESG goals with corporate ESG goals. The chapter discusses and emphasizes Directives of the European Parliament linked to sustainable finance, The Equator Principles, United Nations Environment Programme Finance Initiative, Collevecchio Declaration, European Union Action plan for sustainable finance, EU taxonomy for sustainable activities, and others related to sustainable reporting, Principles for Responsible Investment (PRI), solvency, and Basel.

Chapter 4 identifies the elements of benefits, as well as the costs and risks of the above strategy for the different beneficiaries, as well as identifies the development trends and their impacts on the creation of added value of organizations and their supply chains. In this chapter, the role of sustainable value in social and environmental aspects for and in the organizations is defined and discussed, as are current key drivers for implementing corporate social responsibility (CSR) in innovative business models.

Chapter 5 defines the role of banks in ensuring sustainability in the corporate sector. The chapter points to both corporate motives for sustainability and, in particular, defines them in relation to banks that need to adapt their business models to the current challenges in sustainability and social responsibility. This chapter discusses the roles of the financial sector in sustainable development and indicates the role of the central bank in this process. At the end of the chapter, we present how banks should run green policies in the corporate sector.

Chapter 6 defines sustainable business models and presents areas and frameworks of these innovative models. In this chapter, we present the taxonomy of environmentally friendly business models as well archetypes of these models. The tools for implementation of the sustainable business models in praxis are presented and discussed. The indicated models, tools, and components should provide help to managers reconfiguring the company's strategy.

Chapter 7 identifies what ESG factors have been incorporated by banks in the risk assessment process? In this context, sustainable business models archetypes in the banking sector are discussed, and the challenges and prospects faced by banks when designing their business models toward sustainability and sustainable value creation are identified.

Chapter 8 What does a change in the business model of a financial institution—a bank, mean in the practice of enterprises? Why is it important for companies to take responsibility for their actions, especially today? Why is this responsibility and pro-environmental behavior of enterprises becoming an important decision-making aspect for the bank? These are the questions the chapter responses and the issues that determine the spheres of influence bank—enterprise in sustainable business models.

The book is a source of knowledge on the sustainable business models from the perspective of banks and companies and is dedicated to all those interested in information linked to this scope of interest, in particular to the researchers and practitioners dealing with the aspects and problems of sustainable business models.

CHAPTER 2

Sustainability Impact on Business Models

The main purpose of this study is to show why the sustainability approach is important for building business models. In the chapter, methods to effectively build business models in banks and the approach to redefining them based on risk factors were analyzed. In this section, the authors showed how banks should approach the construction of their business models taking into account their internal specificity, creating original values, the factor of social responsibility, but also ESG risk factors. The end result of this chapter is to answer the question regarding the differences in designing business model canvases and how to approach sustainable business models.

2.1 Introduction

As the impact of sustainable development on the world's future is better understood, each industry is encouraged to find a solution for sustainability in its own field. In the literature on the subject, there is a growing interest in the sustainable finance involving a need to understand and develop ways to take advantage of the potential benefits associated with this emerging way of doing business in the conditions of sustainable development and social responsibility. Many authors see the success of enterprises, financial institutions, or banks in business models. However,

© The Author(s), under exclusive license to Springer Nature Switzerland AG 2021

M. Ziolo et al., *Sustainability in Bank and Corporate Business Models*, Palgrave Studies in Impact Finance,
https://doi.org/10.1007/978-3-030-72098-8_2

this approach or introducing the idea of sustainability to business (Campbell, 1996; Zott & Amit, 2010; Gerster, 2012; Čihák et al., 2012; Saebi & Foss, 2014) is nothing new, for example: how to "do business" (Zott et al., 2011) and how to create financial value (Teece, 2010; Wirtz et al., 2016), how to integrate into one whole basic elements such as systemic analysis, planning, mapping communication as well as configuration (Doleski, 2015; Knyphausen-Aufsess & Meinhardt, 2002), and how to use to create value a strategic asset for competitive advantage and performance of company (Afuah, 2004; Casadesus-Masanell & Ricart, 2010; Chesbrough, 2007; Geissdoerfer et al., 2018).

There is an increasing search for new elements of value, especially financial value. This marks a move away from traditional business models and values. An additional aspect to be taken into consideration when developing a business model is its assessment by stakeholders and society. A change in a business model often effects the social, environmental, and economic performance of the company, institution, or even a bank. Business models and sustainability are not mutually exclusive, but they are two different concepts that work in parallel and even interpenetrate. A new quality emerges from the combination of these concepts, allowing the creation of a new value based on the effects of the social, environmental, and economic. The practice sees the benefits of using sustainability in its activities, and CSR concepts are also important (Lozano, 2014; Roome & Louche, 2015; Joyce & Paquin, 2016; Cardeal et al., 2020). However, few studies exist to support the development of sustainable business models (Holtström et al., 2019).

Several approaches have been developed to conceptualize business models, doing so by dividing them into different elements, but the main concept is based on the canvas concepts (Osterwalder & Pigneur, 2002). It is the basis of the implementation of sustainability to the concept of the business model canvas. There are also important issues of how this concept can develop in financial institutions and how to modify the values of these institutions, especially banks. Practice shows the importance, and even the search for new solutions in building business models of enterprises, in particular financial institutions in the field of sustainability. Responsibility for the environment, the need and challenges related to sustainability, and even openness to Sustainable Development Goals (www. 1), force the inclusion of social responsibility problems in business models and openness to sustainability issues. A new value is created, but there is also an ESG risk that accompanies the implementation of sustainability development naked. Financial institutions, including banks, try to keep up with the business models of their clients. Which means opening up more to sustainability. As contemporary reality shows, and especially the problems of the COVID-19 pandemic, there is a need to take into account the ESG figures. Speech business models must take into account business, environmental, and social aspects. The new approach to the construction of the business model should be designed so that the business can be effective both in the face of the economic crisis and the experience of the COVID-19 pandemic. This applies to both enterprises and financial institutions, their stability and resilience to crises. It is about both sustainability and stability.

The literature lacks approaches to linking these issues, especially in the face of a pandemic. It is necessary to discuss them on the verge of cooperation between enterprises and financial institutions. In order to be able to understand the elements connecting business models toward the sustainability of banks and enterprises, it is necessary to know the factors that will determine the transition toward sustainability, both from the point of view of enterprises and banks. It is important to understand the evolution and specifics of building business models so as to build on the basis of relationships banks (financial institutions)—enterprise interfaces toward sustainability.

2.2 Business Models Versus Sustainability

In recent years, sustainability has been at the core of development in all industries. The idea of sustainability finance stems from the concept of sustainable development, which has been defined as "development that meets the needs of the present without compromising the ability of future generations to meet their own needs" (World Commission on Environment and Development, 1987). When assessing the role and commitment of the financial sector in the economy, this sector plays a key role in the implementation of Sustainable Development Goals (Jeucken, 2001). The financial sector connects all the pillars of sustainable development with its products and services, but also takes actions aimed at forcing the behavior of financed entities to be consistent with Sustainable Development Goals. Effectively operating financial markets ensure efficient capital transfer in the economy, reducing financial risk and ensuring the stable financing of the real economy (Arestis, 2006).

The challenges of sustainable development have necessitated turning the existing order of functioning of paradigms upside down and considering the financial system and finance from the perspective of social, governance, and ecological constraints (and not the other way around) to respond to the challenges of the twenty-first century related to the implementation of the UN Sustainable Development Goals (SDGs). Under the 17 SDGs, 169 specific targets demonstrate the scale of this universal agenda that is aimed to achieve a balance between the economic, social, and environmental aspects of sustainable development (www. 1).

This implies breaking away from financialized technical culture, which tends to see everything from the point of view of an investor (Chiapello, 2018; Lagoarde-Segot, 2020). To achieve sustainable development, an entire business analysis is necessary (Abdelkafi & Täuscher, 2015). The key challenge is to create "economic success through (…] a certain environmental or social activity" (Schaltegger et al., 2012, p. 98). Changes in the perception of sustainable development in creating sustainable finance are presented in Fig. 2.1.

The analogy of three pillars is often criticized because of the unequal consideration, which means if one pillar breaks, it does not impact the entire house (Dawe & Ryan, 2003). Figure 2.2 depicts the idea of three pillars creating sustainable development against the background of financing system. According to Campbell (1996), these issues of sustainability result in three conflicts: (I) economy growth, (II) fair distribution of the growth, and (III) not degrading the ecosystem in the process. Decisions and ratings of current situations are required, especially decisions regarding changes in existing business models, which result in decisions related to the change in the structure of financing and the sources of funding (including obtaining new or alternative funding).

Sustainability necessitates a search for opportunities followed by innovations that are driven by challenges that provoke calls for sustainable development and limiting ESG risks (Hall & Vredenburg, 2003; Hart & Milstein, 2003; Louche et al., 2010; Roome & Louche, 2015).

Sustainability does not detract from business goals, and infusing your company with purpose can help attract a motivated, skilled workforce that drives financial success (Sakis et al., 2016). Sustainability also allows you to attract investors, capital, and potential customers. Thus, new business networks are created around the idea of sustainability, as shown in Fig. 2.1. Sustainability does not mean sacrificing profits, reducing efficiency, losing investors (capital), or putting success on the backburner.

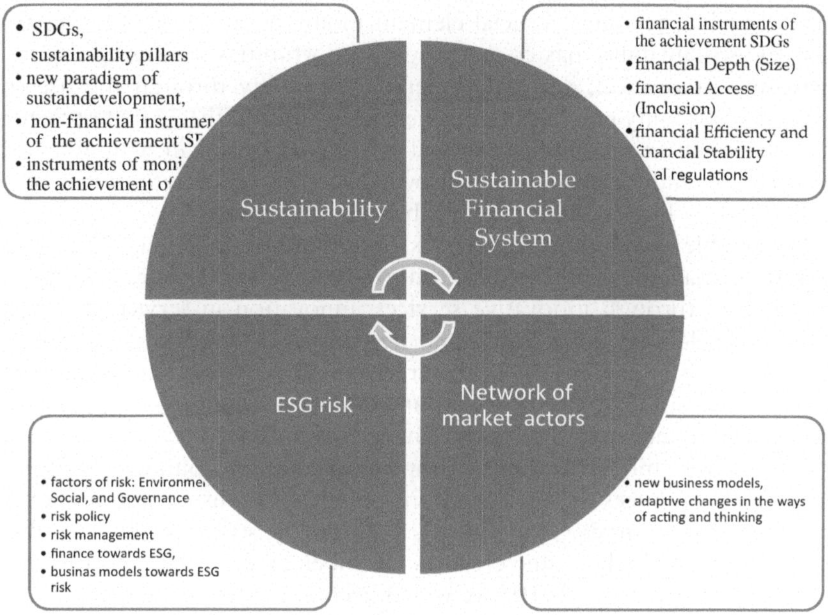

Fig. 2.1 Changes in the perception of sustainable development creating sustainable finance (*Source* Own elaboration based on Gerster [2012], Čihák et al. [2012], Ziolo et al. [2019])

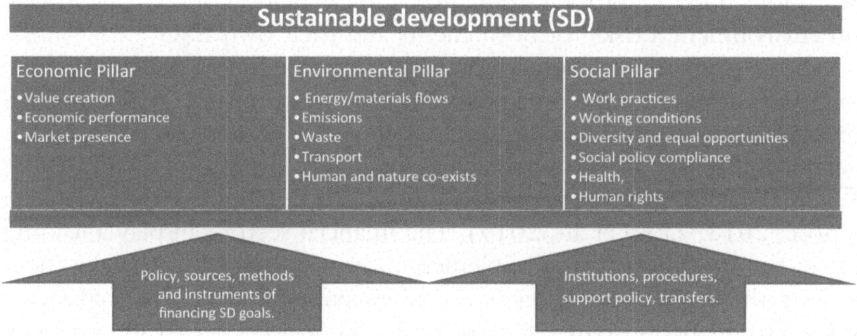

Fig. 2.2 The idea of three pillars in creating sustainable development against the background of the need to finance it (*Source* Own elaboration based on Dawe and Ryan [2003], Holgado Granados [2014], Alińska et al. [2018], Ziolo et al. [2019])

Instead, it has become a crucial element in any organization's successful strategy: sustainability has created wider opportunities for raising capital, increasing innovation, and strengthening credibility through the use of social responsibility practices. A business or institution that does not factor in sustainability risks (ESG risk) will be less successful in a number of measures, including profitability, growth, and employee retention (Sakis et al., 2016; Chladek, 2019; Alsayegh et al., 2020).

Sustainable development provides a competitive advantage but also requires management of the transition to sustainability. This transition can be achieved through innovative services, innovation in service creation, and service delivery approaches, and new forms of business partnerships (Freudenreich et al., 2019). Both directions of activities—managing the transition to sustainability and the pursuit of new solutions—require new business strategies and the use of financial resources (e.g., investments, solutions, and modernizations). Implementation of solutions favoring sustainability reduces the cost to nature and increases sustainable economy capacities (Olawumi & Chan, 2018). Accordingly, strategies that are able to promote sustainable value creation (Dembek et al., 2018) and the cost leadership (Khan et al., 2019) are widely recommended for service sectors to transition toward sustainability (Nosratabadi et al., 2020). As its clients adjust, the financial system must adapt. It must be remembered that financial institutions, in particular banks, always encounter risk, which means that only financial risks are evaluated and calculated. As such, financial institutions must adapt their evaluation and calculation of financial, social, and environmental risks for both clients and their own ESG.

Various studies have been conducted to explore the sustainability practices and strategies of companies and the relationship between such strategies and financial aspects, such as financial performance indicators, the possibility of raising capital, and financing innovation (Mill, 2006; Hess & Warren, 2008; Porter & Kramer, 2011; Gupta, 2017; Mahajan & Bose, 2018; Ziolo et al., 2019). The financial sector can play a leading role in the transition to the implementation of Sustainable Development Goals in the business activities of economic entities. If the financial sector chooses to finance sustainable companies and projects, they can accelerate the transition (Yip & Bocken, 2018). Financial institutions are considered a significant stakeholder in terms of potential to contribute to the SDGs given the resources at their disposal in the form of capital, their sphere of influence in the form of clients and business models, and business opportunities associated with financing Sustainable Development Goals.

The literature on the subject expresses the view that sustainable banks provide their customers with sustainable products that encourage socially responsible investment (Marcel, 2001; Nosratabadi et al., 2020). In addition, the stakeholders judge banks according to their social performance (in particular through the prism of social responsibility and information obtained from Global Reporting Initiative [GRI] reports) and their economic performance (Filipiak & Dylewski, 2020). Financial institutions, and financial markets in particular, award policies that are socially and environmentally responsible (Nosratabadi et al., 2020). Therefore, financial institutions, in particular banks, are starting to change their business strategies, stimulating their clients to focus more on sustainable services and products through redefining their business logic and integrating sustainability in their policies and business activities. The financial system, especially financial institutions, in relation to supporting the 17 SDGs, and the 169 specific targets in particular, must quickly redefine their business models. However, they also have to redefine mutual relations and the principles of cooperation. They need to join forces to assess ESG risk for their business strategies to succeed. These activities must meet the requirements set by the environment (in particular, customers and stakeholders), in cooperation with the public financial system, and with responsible management.

2.3 The Sustainable Business Models Concept

The business model reflects the operating philosophy of the organization (company, institution) and describes in a simplified manner all the elements of the market environment and this entity, as well as the relations between them, which are important for achieving goals and creating specific value for the client (Saebi & Foss, 2014; Zott & Amit, 2010; Santos et al., 2009; Zott & Amit, 2008). Lüdeke-Freund (2010) indicated that the business model describes a competitive advantage through superior customer value and contributes to the sustainable development of companies and society.

The business model concept was originally used to communicate complex business ideas to potential investors within a short time frame (Zott et al., 2011), which means that the business model is used as a unit of analysis offering a systemic perspective on how to do business (Zott et al., 2011), mainly with the aim of understanding how to improve the ability of companies to create financial value (Wirtz et al., 2016) and how to create value (Teece, 2010). The purpose of the concept developed to be currently seen as a tool for the systemic analysis, planning, and mapping communication; for the configuration and implementation of all elements into one whole (Doleski, 2015; Knyphausen-Aufsess & Meinhardt, 2002); and as a strategic asset for competitive advantage and firm performance (Afuah, 2004; Casadesus-Masanell & Ricart, 2010; Chesbrough, 2007; Geissdoerfer et al., 2018). Definitions of a business model often focus on the structural aspects regarding its content (e.g., Johnson et al., 2008; Tikkanen et al., 2005; Voelpel et al., 2005). The business model is combined from different modules that can include alternative actors and roles and value exchanges between them (Palo & Tähtinen, 2014).

The concept of a business model is based on value, but in the literature on the subject, special attention is paid to the ability to create and deliver this value to customers. The value-based view of the business model explains "how a business creates and delivers value, both for the customer and the company" (Johnson, 2010, p. 22) or as "the way individuals or organizations communicate, create, deliver, and capture value out of a value proposition" (Abdelkafi et al., 2013, p. 13). Despite some differences in the constituent components of business models, the original value proposition offered to customers and partners is the central tenet in most of them. In addition to original component value, five main elements of business models are indicated in the literature: value proposition, value network, value capture, value creation, and value delivery (Bocken et al., 2014, pp. 42–56; Boons & Lüdeke-Freund, 2013, pp. 9–19; Palo & Tähtinen, 2014). The changes in the understanding of the concept of value provided to the customer in business models are presented in Fig. 2.3.

When extending the value thread, the PERFA model (P, performance; E, ease of use; R, reliability; F, flexibility; A, affectivity) indicates five important aspects of value for the customer; a value-focused approach requires managers to rethink their perspective on product (service), considering it from the customer's point of view. In doing so, managers are able to identify key factors among the five perspectives of PERFA and make better decisions when deciding what to innovate to improve the value proposition for their customers (Lindič & Marques da Silva 2011; Chambers & Patrocínio, 2011).

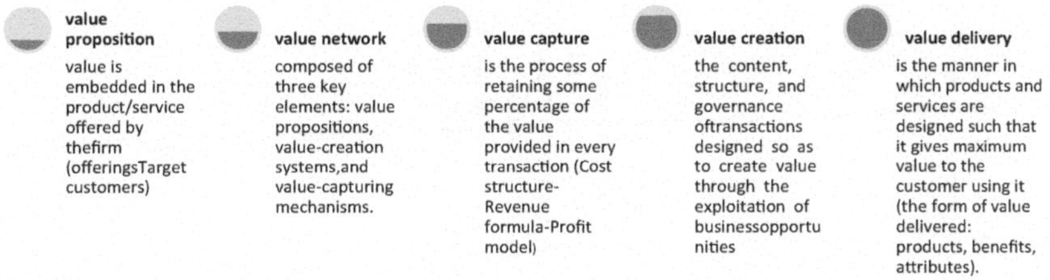

Fig. 2.3 Changes in the understanding of the concept of value provided to the customer in business models (*Source* Own elaboration based on: Guo et al. [2013], Boons and Lüdeke-Freund [2013], Bocken et al. [2014])

In addition to the presence of various definitions of the concept, consensus is lacking on the key components of a business model (Morris et al., 2005). Table 2.1 presents selected grouping concepts of business models components. Business models are understood as manifestations of how organizational variables are configured, how the company structures its transactions with external stakeholders, and the consequences of those configurations on company performance (Saebi & Foss, 2014).

Table 2.1 shows also the various and somewhat overlapping elements of business models that have been presented by different authors. These elements were identified based on a combination of business model literature in search of general components. In addition to the constants, which, after analysis, describe:

- the area of value creation where the company generates value for the customer, most often at the product or service level;
- the area of building relationships with the client, influencing customer satisfaction, and building their commitment;
- the area of infrastructure and resources, enabling the delivery of value to customers;
- the financial area, related to the other three areas in terms of the structure of costs and revenues from customer relations;
- channels enabling communication and delivering value to clients; and
- key operational activities.

Table 2.1 Elements of business models according to selected concepts appearing in literature review

Author(s)	*Years*	***Three and less** categories of business model components*	***Four** categories of model components*	***Six** categories of business model components*	***Seven and more** categories of business model components*
Hamel	2000	–	Basic strategy, strategic resources, value network, customer interface	–	–
Mahadevan	2000	Value stream, logistical stream, and revenue stream	–	–	–
Petrovic, Kittl, and Teksten	2001	–	–	–	Value model, resource model, production model, customer relationship model, revenue model, capital model, market model
Chesbrough and Rosenbloom	2002	–	–	Value proposition, target markets, internal value chain structure, cost structure and profit model, value network, and competitive strategy	–

Author(s)	*Years*	***Three and less*** *categories of business model components*	***Four*** *categories of model components*	***Six*** *categories of business model components*	***Seven and more*** *categories of business model components*
Hedman and Kalling	2003	–	–	–	Customers, competitors, offering, activities and organization, resources, suppliers of factor and production inputs, and scope of management
Voelpel, Leibold, and Tekie	2004	New customer value proposition (a value network (re)configuration, and leadership capabilities	–	–	–
Morris, Schindehutte, and Allen	2005	–	–	Product offering, market factors, internal capability factors, competitive strategy factors, economic factors, and growth/exit factors	–
Osterwalder, Pigneur, and Tucci	2005	–	Product, customer interface infrastructure management, financial aspects,	–	–
Shafer, Smith, and Linder	2005	–	Strategic, value networks, creating value, capturing value	–	–

(continued)

Table 2.1 (continued)

Author(s)	*Years*	***Three and less*** *categories of business model components*	***Four*** *categories of model components*	***Six*** *categories of business model components*	***Seven and more*** *categories of business model components*
Chesbrough	2007	–	–	Value proposition, target market, value chain, revenue mechanism(s), value network or ecosystem, competitive strategy	–
Johnson, Christensen, and Kagermann	2008	–	Customer value proposition—CVP; Key resources, key processes, profit formula	–	–
Richardson	2008	Value proposition, value creation and delivery system; value capture	–	–	–
Demil and Lecocq	2010	Resources and competences; organizational structure, propositions for value delivery	–	–	–
Kujala, Artto, Aaltonen, and Turkulainen	2010	–	–	Customer, value proposition for the customer, competitive strategy, position in the value network, suppliers' internal organization/key capabilities, logic of revenue generation	–

Author(s)	*Years*	***Three and less** categories of business model components*	***Four** categories of model components*	***Six** categories of business model components*	***Seven and more** categories of business model components*
Osterwalder and Pigneur	2010	–	–	–	Customer segments, customer relationships, value propositions, channels, key activities, key resources, key partners, cost structure, revenue streams
Zott and Amit	2010	Design elements, design themes	–	–	–
AI-Debei and Avison	2010		Value proposition, value architecture, value network, finance		
Onetti, Zucchella, Jones, and McDougall-Covin	2012	Focus/what? (activities, resources), (2) locus/where? (location), and (3) modus/how? (internal organization, network design)	–	–	–
Runfola, Rosati, and Guercini	2013	Target segments, value proposition, revenue model	–	–	–
Daft and Albers	2013	Key logic of the company, value chain, assets			

(continued)

Table 2.1 (continued)

Author(s)	*Years*	***Three and less** categories of business model components*	***Four** categories of model components*	***Six** categories of business model components*	***Seven and more** categories of business model components*
Bocken, Short, Rana, and Evans	2014	Value proposition; value creation and delivery value capture	–	–	–
Abdelkafi and Täuscher	2016	Value proposition, value creation, value capture	–	–	–
Roome and Louche	2016	–	Value proposition, value network, value capture, value creation and delivery	–	–
Kaplan and Winby	2019	Organization creates, delivers and captures value			
Ramdani, Binsaif, and Boukrami	2019		Value proposition, operational value, human capital, financial value		
Summary		11	7	4	3

Source Own elaboration

There are elements that distinguish business strategies, which depend on the market, industry, company size, respect for the value of sustainability, and social responsibility. In the literature on the subject, two groups of elements describing the business model are indicated (Palo & Tähtinen, 2014): firstly, some elements describe the business model from a firm's perspective, which are underlaid, secondly, by elements that describe the business model from a business net perspective of the actors and their roles, the specific value system, relations and ties, and dynamics. Theoretical discussion showed that business nets are an essential element of business models; thus, the two theoretical premises are strongly interconnected (Palo & Tähtinen, 2014; Laasch, 2019).

For the development of the concept of business models, Zott and Amit (2010) defined the relationship between the components of the business model they recognized and the sources of value creation. They referred to it as the "general framework of the business model" (Table 2.2). Amit and Zott mentioned elements (design elements) and motives of the business model. The elements include content, structure, and supervision of the business model (governance). The motives of the business model (design themes) are novelty, lock-in (maintenance), complementarity, and efficiency.

Table 2.2 The general framework of the business model

The elements of the business model	*The motives of the business model*
Design elements	
Content	What tasks should be performed?
Structure	How should the tasks be related and in what order should they be performed?
Governance	Who and where should perform the tasks? How to supervise the implementation of tasks?
Design themes	
Novelty	Adopt innovative content (novelity), structure or surveillance items
Lock-in (maintenance)	Building conditions enabling the maintenance of the current business model stakeholders
Complementarity	Mutual complementarity of products or services
Efficiency	Redesigning activities to reduce operating costs

Source Own elaboration based on: Zott and Amit (2010)

Recent literature has widely discussed the elements of business models (Table 2.1) and has paid particular attention to the concepts of configuration (Baden-Fuller & Mangematin, 2013), purposes (Baden-Fuller & Morgan, 2010), customer value creation and value capture (Amit & Zott, 2001; Teece, 2010), design (Zott & Amit, 2007, 2010), and interaction with technology (Baden-Fuller & Haefliger, 2013; Chesbrough & Rosenbloom, 2002; Chesbrough, 2007). However, a clear answer to the question of how companies change and develop their business model remains elusive (Achtenhagen et al., 2013). Scientific works have sought to understand the evolution and dynamism underscoring new business models (Achtenhagen et al., 2013; Cavalcante et al., 2011; Demil & Lecocq, 2010; Sabatier et al., 2010; Roome & Louche, 2015), and the transformational approach considers the business model as a concept or tool that is used to address change with a focus on innovation (Demil & Lecocq, 2010; Roome & Louche, 2015). Directing business models to sustainability and ESG risk is strongly emphasized (Roome & Louche, 2015).

2.4 Developing Business Models: Toward a Unified Perspective for Creation of Sustainable Business Models

In the literature on the subject, it is assumed that the term *business model* was used for the first time in a publication in 1957, which concerned the creation of a mathematical model with the intention of using it in a business game (Bellman et al., 1957). The term was used to present the business reality, and the development of a model with its use consisted of the implementation of specific procedures resulting from the creation of a mathematical model and the performance of the simulation process based on the created model.

Morris et al. (2005) analyzed the types of decisions and distinguished three perspectives in business models: financial, operational, and strategic. Each of the three types of perspectives is related to a different area of decisions made in companies. The financial perspective defines the business model as a financial model that shows how the company generates profits and builds its effectiveness. In financial terms, it describes the operational and strategic aspects, considering the structure of costs and revenues (Teece, 2010). The operational perspective emphasizes the internal processes that enable the company to create value, and the decisions relate to the architecture and configuration of these processes. This approach is related to the current decisions. The strategic perspective indicates aspects related to the company's positioning, cooperation with other entities and financial institutions, and its development opportunities, which are related to decisions regarding stakeholder identification, value creation, differentiation, vision, values, networks, and alliances (Morris et al., 2005).

In the literature on the subject, the research stream devoted to business models concerning e-business can be distinguished. This business model focuses on doing business on the Internet (or by electronic means) and refers to the concept of values, financial aspects, network functioning, and cooperation with other entities (Zott et al., 2011). Rapid advances in communication and information technologies, such as the emergence and swift expansion of the Internet, easy access to information technology, and the rapid decline in computing and communication costs, have allowed the development of new methods to create and deliver value, offering scope for the creation of unconventional exchange mechanisms and transaction architectures, including using websites to display information for products or services, accentuating the possibilities for the design of new boundary-spanning organizational forms (Dunbar & Starbuck, 2006; Zott et al., 2011). The e-business model includes the roles and relationships among a firm's customers, allies, and suppliers. It also includes the major flows of products, services, information, various financial aspects, and the major benefits to the participants. The model includes eight ingredients of business: value proposition, revenue model, market opportunity, competitive environment, competitive advantage, market strategy, organizational development, and management team (Aithal, 2016). This concept was developed by e-commerce business models; in particular, it enabled relationships such as business-to-customer (B2C), business-to-business (B2B), customer-to-customer (C2C), and even customer-to-business (C2B) (D'Andrea et al., 2014).

A look at the business model from the perspective of innovation management was presented by Chesbrough and Rosenbloom (2002), according to whom the business model is a heuristic logic combining technological potential with the realization of economic value. In this concept, the business model is mainly seen as a mechanism that connects a firm's (innovative) technology to customer needs and/or to other company resources. This model is conceptually placed between a firm's input resources and market outcomes, and it "embodies nothing less than the organizational and financial 'architecture' of the business" (Teece, 2010, p. 173). Firstly, companies commercialize innovative ideas and technologies through their business models. Secondly, the business model represents a new subject of innovation, which complements the traditional subjects of process, product, and organizational innovation and involves new forms of cooperation and collaboration (Foss & Saebi, 2015). Business models not only have consequences for technological innovations but can also be shaped by them (Zott et al., 2011). Business model innovation can be a pathway to competitive advantage if the model is sufficiently differentiated and hard to replicate for incumbents and new entrants alike. The business model, according to this more functionalist perspective, complements technology, but technology is seen as an enabler of the business model rather than as part of the concept per se. Neither input resources nor competition in output markets are considered part of the business model concept (Zott et al., 2011). Business model innovation can be classified through the prism of the three dimensions of innovation in technology, value network, and financial hurdle rate. Giesen et al. (2007) and Lindgardt et al. (2009) conceptualized business model innovation as innovations in value proposition, revenue model, and operating model. The present literature on business model innovation mainly focuses on external antecedents, which may pressure companies to engage in business model innovation (Foss & Saebi, 2017; Lee et al., 2019).

In the literature, research on business models from the perspective strategic issues has revolved mainly around three aspects: (1) the networked nature of value creation, (2) the relationship between business models and firm performance, and (3) the distinction between the business model and other strategy concepts (Zott et al., 2011). The business model may articulate the logic, the data, and other evidence supporting the delivery of the value proposition to customers, revenue and cost models, and partners' liabilities. The business model may be ineffective and merely an imitation if not supported by appropriate strategies. Thus, changes in business models or business model innovations are supported by strategies that lead firms to achieve competitive advantages (Mustafa & Werthner, 2011). The business model is understood here as a system of interconnected and interdependent activities defining the way of doing business in cooperation with clients, partners, and suppliers (Zott & Amit, 2010). The system of activities goes beyond the boundaries of the company, and its design requires consideration of two types of parameters: design (such as scope, structure, and management) and thematic (novelty, customer retention, complementarity, and efficiency). A business model resembles a template that defines how a company conducts business, how it provides value, and how it combines production factors into market products.

Business models need to be created to preserve the delicate balance between strategy, and business model design determines the long-term competitive advantage and profitability of a company. Focusing on strategy without heeding the business model can cause companies to lose sight of changing customer behaviors (Braun et al., 2019).

When defining a business model, many authors have indicated its components or strategic dimensions. The development of this work led to the formulation of four directional components (Chambers & Patrocínio, 2011; Fielt, 2013; Wu & Sivalogathasan, 2013; Šimberová & Kita, 2020):

- Strategic choices: clients, target sector, value proposition, skills, revenues, competitors, offer, strategy, branding, diversification, and mission;
- Creating value: resources and assets, innovations, processes, and activities;
- Taking over of value: costs, profits, financial effects, and financial aspects; and
- Value network: suppliers, customer information, customer relations, information, product, and service flows.

Directional strategic dimensions have been adopted that determine the business model of the company (Morris et al., 2005; Chambers & Patrocínio, 2011; Tanwar, 2013):

- target customer (group) that the company wants to serve (the analysis includes, in particular, the profile of customers and defining opportunities for them to increase the actual value, defining key customers from the point of view of profit and markets, and defining the customer sector in which the company is not interested);
- value model (e.g., how the company makes a profit, how is value created and captured for customers, and its profit model);

- differentiation: strategic control and monitoring (how to protect the profit stream, what makes the company's offer unique, what strategic control points can counterbalance the strength of competitors, and identifying what benchmarks look like);
- the scope of the company's operations (what products, services, and solutions the company wants to sell; which is a force that builds value; which activities or functions to perform within the company and which to outsource to other entities).

Important for the advancement of knowledge and science is the concepts that are useful for practice. Therefore, activities were undertaken to analyze business models and their usefulness for business. The purpose of the analysis of business models is to deepen and broaden the knowledge about the basic components of a business model. We perceive the importance of this aim in improving the functionality and economy of the business models, and in "discovering and developing competitive advantage, which can be detected by the companies themselves" (Slávik & Bednár, 2014). The discussion of this trend was conducted in the following directions:

- The successful business model, per Mullins and Komisar (2009), stands on five pillars that predetermine the economic viability of the business: revenue model, gross margin model, operating model, working capital model, and investment model. This concept can be applied to analyze business economy and evaluate financial health, but it abstracts from other components of the business model.
- Afuah (2004) divided the model into four components: industry factors, resources, resources, and costs. This model emphasizes the determinants of profitability, which influence all the activities of the company. This model does not comprehensively define the company as a complex system.
- Watson (2005) showed and evaluated the business model through six components: competitors, customers, economy, management, products, and suppliers. This model analyzes sector factors, such as competition, which belong to the environment of the business model, but is not part of business model components; therefore, it has a comprehensive nature.
- Johnson et al. (2008) defined the business model as a set of four components, which are interconnected: value for a customer, profit

formula, key resources, and key activities. This concept comprehensively describes all the essential components of business.

The business model canvas (BMC) , was developed by Osterwalder and Pigneur (2009), based on the practical experience of entrepreneurs from all over the world. Business models have also evolved toward specific domain solutions, and so there are models specific to tourism, industry, or banking and financial institutions. The emerging need to live in harmony with the environment and the expanding influence of the idea of sustainability have resulted in the emergence and wide development of business models toward sustainability. The answer to the need to combine the effects of BMC and sustainability is the triple-layer business model canvas (TLBMC). This concept is a practical and easy-to-use tool that supports creatively developing, visualizing, and communicating sustainable business model innovation (Stubbs & Cocklin, 2008). The TLBMC follows a triple-bottom-line approach to organizational sustainability (Elkington, 1994), explicitly addressing and integrating economic, environmental, and social value creation as core to an organization's business model. In particular, it leverages life cycle analysis and stakeholder management perspectives within newly created environmental and social canvases to conceptualize and connect multiple types of value creation within a business model perspective.

2.5 Business Models and Business Model Canvases: Designing Sustainable Business Models

Business models can be understood as a conceptual tool that contains a set of elements and their relationships and allows expressing a company's logic of earning money. It is a description of the value a company offers to one or several segments of customers and the design of the firm and its network of partners for creating, marketing, and delivering this value and relationship capital, in order to generate profitable and sustainable revenue streams (Osterwalder, 2004). According to Osterwalder and Pigneur, the business model consists of nine basic elements that fit into the logic of the company's process of making money (Osterwalder & Pigneur, 2010; Tokarski et al., 2017). These nine elements touch four core areas of business activity: customers, offers, infrastructure, and appropriate financial position. The business model is related to the strategy and can be considered a sketch of the strategy to be implemented within the structures, processes, and systems of an organization (Osterwalder & Pigneur, 2012).

The business model canvas (BMC) is a tool for business model development.

The basis for creating the BMC was the work of Slywotzky (1995) and Markides (1999), who distinguished a few simple pillars, which are the what, who, and how of a business, to which we would add the "how much," which expresses what a company offers, who it targets with the offer, how this can be realized, and how much can be earned by doing it (Osterwalder & Pigneur, 2002).

The BMC visualizes the framework of the different elements of how a business works. It illustrates what the business does, for and with whom, and the resources it needs to achieve that. In practice, the canvas has become the quasi-standard for describing business models (Jones & Gregor, 2006). The canvas is composed of nine different elements. Table 2.3 lists the elements of the business model according to Osterwalder and Pigneur.

Defining these elements allows us to examine the planned business from a completely different perspective. Figure 2.4 presents a diagram of the business model canvas. The diagram facilitates planning and building new solutions, as well as improving those already used in the company. It also allows identification of areas of innovation.

The BMC is seen as essential for entrepreneurs to constantly reflect and to develop their business model (Trimi & Berbegal-Mirabent, 2012; Steen, 2013). When creating a business model, the order in which each element is determined should be appropriate. The model starts with a value proposition and ends with a cost structure. This makes it easier to define subsequent elements. The literature on the subject provides four basic principles of design from the BMC, which, along with the order in which each canvas is specified, should be considered:

Design principle 1: frame the ill-structured problem by developing an ontology in which the main components of the problem and their relationships are modeled. To tackle this question a rigorous conceptual model of business models was developed, which was called the business model ontology (BMO) (Avdiji et al., 2018). The BMO is based on the methodological approach of building ontologies: (1) identifying the key concepts and relationships in the domain of business models, (2) producing precise and unambiguous text definitions for the concepts and

Table 2.3 Elements of the business model BMC according to Osterwalder and Pigneur concept

Business areas	*Canvas*	*Description of the canvas (business model element)*
Offer	Value proposition	It is necessary to determine the answer to the question which problem solves or what customer needs using product or service will satisfy; it is important to determine and show (establishing or estimating) what value it will have for the customer
Customers	Segments	It is definition of the group the company dedicates its product. Defining the profile of your ideal client, as well as individual customer segments with own characteristics
	Relationships	It presents the characteristics of the relationship between the company and all clients from a specific segment
	Channels	It indicates how the company communicates with the individual segments of its customers and how it communicates its value proposition to them
Infrastructure	Key resources	It indicates the most important resources necessary for the proper functioning of the business model
	Key actions	It presents the most important activities that the company needs to undertake so that its model functions efficiently and brings the intended goals
	Key partners	It describes the network of suppliers and associates on whom the smooth functioning of the company depends
Financial situation	Revenue streams	It symbolizes the amount of funds generated by company in connection with servicing each of the segments, financial flow plans and generated profits
	Cost structures	Estimated fixed and variable costs of the business model adopted and identify which aspects of the business are the costliest, cost models

Source Own elaboration on Osterwalder and Pigneur (2002), Osterwalder et al. (2005), Jones and Gregor (2006), Osterwalder and Pigneur (2010), Fielt (2013), Qastharin (2015)

Key Partners

WHO:
- Who are our Key Partners?
- Who are our key suppliers?

WHICH:
- Which Key Resources are we acquiring from partners?
- Which Key Activities do partners perform?

MOTIVATIONS FOR PARTNERSHIPS:
- Optimization and economy;
- Reduction of risk and uncertainty;
- Acquisition of particular resources and activities

Key Activities

KEY ISSUES:
- What Key Activities do our Value Propositions require?
- Our Distribution Channels?
- Customer Relationships? Revenue streams?

CATEGORIES:
- Production,
- Problem Solving,
- Platform/Network

Key Resources

KEY ISSUES:
- What Key Resources do our Value Propositions require?
- Our Distribution Channels?
- Customer Relationships Revenue Streams?

TYPES OF RESOURCES:
- Physical,
- Intellectual (brand patents, copyrights, data),
- Human,
- Financial

Value Propositions

KEY ISSUES:
- What value do we deliver to the customer?
- Which one of our customer's problems are we helping to solve?
- What bundles of products and services are we offering to each Customer Segment?
- Which customer needs are we satisfying?

CHARACTERISTICS:
- Newness,
- Performance,
- Customization,
- "Getting the Job Done",
- Design, Brand/Status,
- Price,
- Cost Reduction,
- Risk Reduction,
- Accessibility,
- Convenience/Usability

Customer Relationships

KEY ISSUES:
- What type of relationship does each of our Customer Segments expect us to establish and maintain with them?
- Which ones have we established?
- How are they integrated with the rest of our business model?
- How costly are they?

Channels

WHICH:
- Through which Channels do our Customer Segments want to be reached?
- Which ones work best?
- Which ones are most cost-efficient?

HOW:
- How are we reaching them now?
- How are our Channels integrated?
- How are we integrating them with customer routines?

Segments

WHO:
- For whom are we creating value?
- Who are our most important customers?

KEY ISSUES:
Is our customer base a Mass Market, Niche Market, Segmented, Diversified, Multi-sided Platform

Cost Structure

KEY ISSUES:
- What are the most important costs inherent in our business model?
- Which Key Resources are most expensive?
- Which Key Activities are most expensive?

KEY PROBLEMS:
- Cost Driven (leanest cost structure, low price value proposition, maximum automation, extensive outsourcing),
- Value Driven (focused on value creation, premium value proposition).

KEYS CHARACTERISTICS:
- Fixed Costs (salaries, rents, utilities),
- Variable costs,
- Economies of scale,
- Economies of scope,
- Cost models.

Revenue Stream

KEY ISSUES:
- For what value are our customers really willing to pay?
- For what do they currently pay?
- How are they currently paying?
- How would they prefer to pay?
- How much does each Revenue Stream contribute to overall revenues?

TYPES:
- Asset sale,
- Usage fee,
- Subscription Fees,
- Lending/Renting/Leasing, Licensing, Brokerage fees, Advertising

FIXED PRICING:
- List Price,
- Product feature dependent,
- Customer segment dependent,
- Volume dependent

DYNAMIC PRICING:
- Negotiation (bargaining),
- Yield Management,
- Real-time-Marke

Fig. 2.4 Diagram of business model canvas (*Source* Own elaboration on Osterwalder and Pigneur [2002, 2010], Osterwalder et al. [2005])

their relationships, and (3) identifying the terms to refer to these concepts and relationships using labels (Uschold & King, 1995; Osterwalder & Pigneur, 2002; Avdiji et al., 2018). Figure 2.5 shows a representation of the different elements and how they relate to each other, which is called the business model ontology schematic.

Design principle 2: represent the ontology as a shared visualization by deriving a concept map from the ontology and logically structuring the concepts into visually empty problem spaces. In response to this design principle, a schematic tool called the map of business model canvas (Fig. 2.5) was developed. A conceptual map was built upon the ontology to relate all the concepts (Osterwalder & Pigneur, 2010). Design principle 2 is based on the assumption that (1) the business model tool should help business practitioners more quickly understand a business model and the relationships behind its elements, and (2) it should create a common language to improve communication between the stakeholders when addressing business model issues (Osterwalder & Pigneur, 2013; Avdiji et al., 2018).

Design principle 3: use visual methods to facilitate the work of business teams; and create, manipulate, assess, and discuss such objects to create new strategic alternatives. This principle is about dividing teams

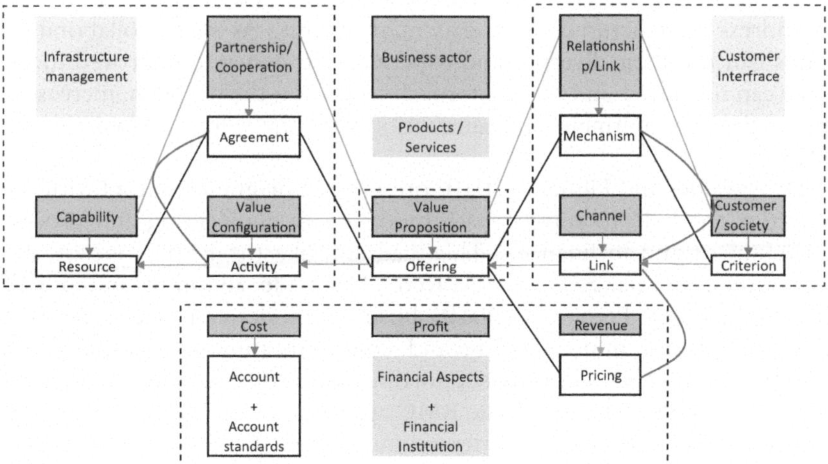

Fig. 2.5 Schematic business model canvas ontology (*Source* Own elaboration on Osterwalder and Pigneur [2002], Avdiji et al. [2018])

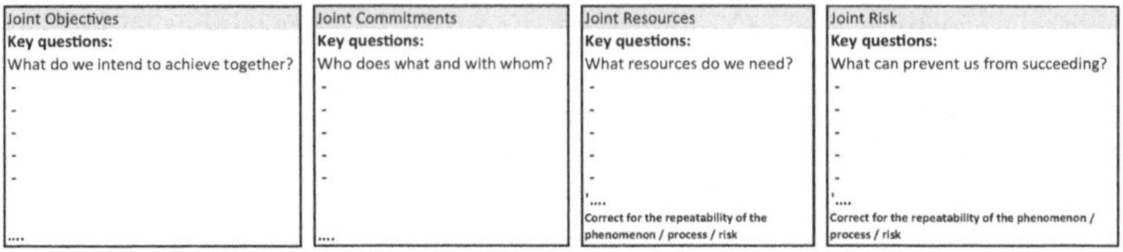

Fig. 2.6 Schematic idea of the team alignment map (*Source* Own elaboration on Osterwalder and Pigneur [2002, 2013], Avdiji et al. [2018])

based on specialization, as well as about bringing them together. The team alignment map may be used for this purpose. This map helps with achieving goals on time while dramatically reducing miscommunication, frustration, and unnecessary stress. It is vital to make objectives and objects more tangible using visual representations. This allows a group of business designers to easily and jointly sketch, create, manipulate, assess, and discuss such objectives to create new strategic alternatives using a common language (Osterwalder & Pigneur, 2013; Avdiji et al., 2018). A scheme of the team alignment map is shown in Fig. 2.6.

Design principle 4: instantiate the visualization into a shared support to use it as a problem space in which solutions can be prototyped. Cooperation, exchange of experiences, and ideas of team members produce results to address ill-structured management problems. As such, collaborating teams can use their diversity and experience to generate improved ideas. They can use a trial-and-error approach to solving the problem, increasing the level of commitment and satisfaction with the solution (Avdiji et al., 2018).

Osterwalder and Pigneur (2010) presented a diagram and tool with the objective to "help to support task modeling in engineering and process management and in finance." This diagram, like the BMC concept, has been updated, changed, and evolved. The wide adoption and use of Osterwalder and Pigneur's (2010) business model canvas suggest that it is valuable for understanding and communicating an organization's business model and creatively supporting business innovation (Abraham, 2013). Unfortunately, the classic BMC approach is insufficient for organizations seeking sustainability-oriented innovation (Lozano, 2014). Therefore, new creative tools are needed. There is a need to include new environmental and social layers in the original business model canvas (Osterwalder & Pigneur, 2010), which will be the focus of the next section as part of a wider aspect. The triple-layer business model canvas (TLBMC) enables a triple bottom line perspective to sustainability—that of economic, environmental, and social impact—to be applied in a business model. The TLBMC expands the economic-centered approach to a standard business model by developing and integrating environmental and social canvas layers built from lifecycle and stakeholder perspectives into an extended business model canvas (Joyce & Paquin, 2016).

2.6 Conclusion

Business models give undoubtedly benefits for organizations and allow you to create value for the client and the organization itself. It is problematic and time-consuming to move from traditional business models to those based toward sustainability. However, but these are not easy complex business models to implement. However, there are many tools and concepts that facilitate and demonstrate the best solutions that organizations can use to implement such innovative solutions for the environment and society. Business models help to create value for the organization, and the identified driving forces allow for cost savings, better financial performance, or a better image of the organization on the market.

The spread of the COVID-19 pandemic in 2020 also affected the realization of SDGs in organizations and institutions. In the context of CSR areas, it can be pointed out that the pandemic affected (García-Sánchez & García-Sánchez, 2020) the level of economic risk at the business level. This situation made it necessary to verify not only the existing business models, but also made it necessary to extend the existing risks taken into account when building business models with ESG risks. A pandemic certainly contributes to redefining all spheres of business, but non-financial factors are gaining in importance. Therefore, there is a need to take into account more non-financial qualitative factors and ESG risks as elements of the sustainability concept appear in business models.

In the case of building new solutions that will force the construction of new business models, it can be expected that the effectiveness of their impact in the context of the results obtained will be unsatisfactory in relation to the expectations declared during their establishment. The reason for the ineffectiveness, or even the adequacy of business models, may be too little share of the sustainability factor or poor recognition and consideration of ESG risks and the pandemic factor. It is therefore important to design business models so that canvas takes into account both sustainability, ESG risks and the pandemic factor.

References

Abdelkafi, N., Makhotin, S., & Posselt, T. (2013). Business model innovation for electric mobility: What can be learned from existing business model patterns? *International Journal of Innovation Management, 17*, 1–41. https://doi.org/10.1142/S1363919613400033.

Abdelkafi, N., & Täuscher, K. (2015). Business models for sustainability from a system dynamics perspective. *Organization & Environment, 29*(1), 74–96.

Abraham, S. (2013). Will business model innovation replace strategic analysis? *Strategy & Leadership, 41*(2), 31–38.

Achtenhagen, L., Melin, L., & Naldi, L. (2013). Dynamics of business models–strategizing, critical capabilities and activities for sustained value creation. *Long Range Planning, 46*(6), 427–442.

Alińska, A., Filipiak, B. Z., & Kosztowniak, A. (2018). The importance of the public sector in sustainable development in Poland. *Sustainability, 10*, 3278.

Afuah, A. (2004). *Business models: A strategic management approach*. Boston, MA: McGraw-Hill, Irwin.

Aithal, P. S. (2016). A review on various E-business and M-business models & research opportunities. *International Journal of Management, IT and Engineering, 6*(1), 275–299.

Alsayegh, M. F., Rahman, R., & Homayoun, S. (2020). Corporate economic, environmental, and social sustainability performance transformation through ESG disclosure. *Sustainability, 12*, 3910. https://doi.org/10.3390/su12093910.

Amit, R., & Zott, C. (2001). Value creation in e-business. *Strategic Management Journal, 22*, 493–520.

Arestis P. (2006). Financial liberalization and the relationship between finance and growth. In P. Arestis & M. Sawyer (Eds.), *A handbook of alternative monetary economics* (pp. 346–364). Edward Elgar.

Avdiji, H., Elikan, D., Missonier, S., & Pigneur, Y. (2018). Designing tools for collectively solving III-structured problems. *HICSS, 2018*, 1–10.

Baden-Fuller, C., & Haefliger, S. (2013). Business models and technological innovation. *Long Range Planning, 46*, 419–426. https://doi.org/10.1016/j.lrp.2013.08.023.

Baden-Fuller, C., & Mangematin, V. (2013). Business models: A challenging agenda. *Strategic Organization, 11*(4), 418–427.

Baden-Fuller, C., & Morgan, M. S. (2010). Business models as models. Long Range Planning, 43(2–3), 156–171. https://doi.org/10.1016/j.lrp.2010.02.005.

Bellman, R., Clark, Ch., Craft, C., Mlcolm, D. G., & Riccardi, F. (1957). *On the construction of a multi-stage, multi-person business game* (pp. 469–503). Operations Research, The Rand Corporation.

Bocken, N. M., Short, S. W., Rana, P., & Evans, S. (2014). A literature and practice review to develop sustainable business model archetypes. *Journal of Cleaner Production, 65,* 42–56.

Boons, F., & Lüdeke-Freund, F. (2013). Business models for sustainable innovation: State of the art and steps towards a research agenda. *Journal of Cleaner Production, 45,* 9–19.

Braun, M., Latham, S., & Cannatelli, B. (2019). Strategy and business models: Why winning companies need both. *Journal of Business Strategy, 40*(5), 39–45.

Campbell, S. (1996). Green cities, growing cities, just cities? Urban planning and the contradictions of sustainable development. *Journal of the American Planning Association, 62*(3), 296–312.

Cardeal, G., Höse, K., Ribeiro, I., & Götze, U. (2020). Sustainable business models-canvas for sustainability, evaluation method, and their application to additive manufacturing in aircraft maintenance. *Sustainability, 12,* 9130. https://doi.org/10.3390/su12219130.

Casadesus-Masanell, R., & Ricart, J. E. (2010). From strategy to business models and onto tactics. *Long Range Planning, 43,* 195–215.

Cavalcante, S., Kesting, P., & Ulhøi, J. (2011). Business model dynamics and innovation: (re)establishing the miss-ing linkages. *Management Decision, 49,* 1327–1342. https://doi.org/10.1108/00251741111163142.

Chambers, E., & Patrocínio, M. (2011). *Business models and value creation: A case study of The New York City economic development corporation.* Umeå School of Business.

Chesbrough, H. W. (2007). Business model innovation: It's not just about technology anymore. *Strategy and Leadership, 35,* 12–17.

Chesbrough, H. W., & Rosenbloom, R. S. (2002). The role of the business model in capturing value from innovation: Evidence from Xerox Corporation's technology spinoff companies. *Industrial and Corporate Change, 11,* 533–534.

Chiapello, E. (2018). The work of financialisation (Chapter 21). In I. Chambost, M. Lenglet, & Y. Tadjeddine (Eds.), *The making of finance: Perspectives from the social sciences.* Oxon, UK: Routledge.

Chladek, N. (2019, November 6). *Why you need sustainability in your business strategy.* Business Insights. https://online.hbs.edu/blog/post/business-sustainability-strategies.

Čihák, M., Demirgüç-Kunt, A., Feyen, E., Levine, R. (2012). Benchmarking financial systems around the world. The World Bank, Financial and Private Sector Development Vice Presidency & Development Economics Vice Presidency. Available online https://openknowledge.worldbank.org/bitstream/handle/10986/12031/wps6175.pdf.

D'Andrea, A., Ferri, F., & Grifoni, P. (2014). The E-commerce business model implementation. In J. Wang (Ed.), *Encyclopedia of business analytics and optimization* (pp. 2509–2520). IGI Global.

Dawe, N. K., & Ryan, K. L. (2003). The faulty thee-legged-stool-model of sustainable development. *Conversation Biology, 24*, 1458–1460.

Dembek, K., York, J., & Singh, P. J. (2018). Creating value for multiple stakeholders: Sustainable business models at the base of the pyramid. *Journal of Cleaner Production, 196*, 1600–1612.

Demil, B., & Lecocq, X. (2010). Business model evolution: In search of dynamic consistency. *Long Range Planning, 43*, 227–246. https://doi.org/10.1016/j.lrp.2010.02.004.

Doleski, O. D. (2015). *Integrated business model: Applying the St. Gallen management concept to business models*. Springer Gabler.

Dunbar, R. L. M., & Starbuck, W. H. (2006). Learning to design organizations and learning from designing them. *Organization Science, 17*, 171–178.

Elkington, J. (1994). Towards the sustainable corporation: Win-win strategies for sustainable development. *California Management Review, 36*(2), 90–100.

Fielt, E. (2013). Conceptualising business models: Definitions, frameworks and classifications. *Journal of Business Models, 1*(1), 85–105.

Filipiak, B., & Dylewski, M. (2020). Sustainable financial reporting in the context of ensuring sustainability of financial systems. In M. Ziolo (Ed.), *Finance and sustainable development* (pp. 235–267). Routledge.

Foss, N. J., & Saebi, T. (2015). Business models and business model innovation: Bringing organization into the field. In N. J. Foss & T. Saebi (Eds.), *Business model innovation: The organisational dimension* (pp. 1–23). Oxford University Press.

Foss, N. J., & Saebi, T. (2017). Fifteen Years of research on business model innovation: How far have we come, and where should we go? *Journal of Management, 43*(1), 200–227.

Freudenreich, B., Lüdeke-Freund, F., & Schaltegger, S. (2019). A stakeholder theory perspective on business models: Value creation for sustainability. *Journal of Business Ethics, 40*(1), 98.

García-Sánchez, I.-M., & García-Sánchez, A. (2020). Corporate social responsibility during COVID-19 pandemic. *Journal of Open Innovation: Technology, Market, and Complexity, 6*, 126. https://doi.org/10.3390/joitmc6040126.

Geissdoerfer, M., Vladimirova, D., & Evans, S. (2018). Sustainable business model innovation: A review. *Journal of Cleaner Production, 198*.

Gerster, R. (2012). *Sustainable finance: Achievements, challenges, outlook*. Striking a Balance Ahead of Rio+20in 2012. Available online http://www.gersterconsulting.ch/docs/sustainable_finance_final_11.02.10.pdf.

Giesen, E., Berman, S., Bell, R., & Blitz, A. (2007). Paths to success, three ways to innovate your business model. *Journal of Strategy and Leadership, 35*(6), 27–33.

Guo, H., Zhao, J., & Tang, J. (2013). The role of top managers' human and social capital in business model innovation. *Chinese Management Studies, 7*(3), 447–469.

Gupta, H. (2017). Integration of quality and innovation practices for global sustainability: An empirical study of Indian SMEs. *Global Business Review, 18*(1), 210–225.

Hall, J., & Vredenburg, H. (2003). The challenges of innovating for sustainable development. *MIT Sloan Management Review, 45*(1), 61–68.

Hart, S. L., & Milstein, M. B. (2003). Creating sustainable value. *Academy of Management Executive, 17*(2), 56–69.

Hess, D., & Warren, D. E. (2008). The meaning and meaningfulness of corporate social initiatives. *Business and Society Review, 113*(2), 163–197.

Holgado Granados, M. (2014). Sustainable value creation in manufacturing through maintenance services. Ph.D. thesis, University of Milan.

Holtström, J., Bjellerup, C., & Eriksson, J. (2019). Business model development for sustainable apparel consumption: The case of Houdini Sportswear. *Journal of Strategy and Management, 12*(4), 481–504. https://doi.org/10.1108/JSMA-01-2019-0015.

Jeucken, M. (2001). *Sustainable finance and banking: The financial sector and the future of the planet*. Earthscan.

Johnson, M. W. (2010). *Seizing the white space: Business model innovation for growth and renewal*. Boston: Harvard Business Press.

Johnson, M. W., Christensen, C. M., & Kagermann, H. (2008). Reinventing your business model. *Harvard Business Review, 86*(12), 50–59.

Jones, D., & Gregor, S. (2006). *The formulation of an information systems design theory for e-Learning*. Paper presented at the First International Conference on Design Science Research in Information Systems and Technology, Claremont, CA, 356–373.

Joyce, A., & Paquin, L. R. (2016). The triple layered business model canvas: A tool to design more sustainable business models. *Journal of Cleaner Production, 135*, 1474–1486.

Khan, S. Z., Yang, Q., & Waheed, A. (2019). Investment in intangible resources and capabilities spurs sustainable competitive advantage and firm performance. *Corporate Social Responsibility and Environmental Management, 26*, 285–295.

Knyphausen-Aufsess, D. Z., & Meinhardt, Y. (2002). Revisiting Strategy: ein Ansatz zur Systematisierung von Gesch€aftsmodellen. In T. Bieger, N. Bickhoff, R. Caspers, D. Z. Knyphausen-Aufseß, & K. Reding (Eds.), *Zukünftige Gesch€aftsmodelle*. Springer.

Laasch, O. (2019). An actor-network perspective on business models: How 'Being Responsible' led to incremental but pervasive change. *Long Range Planning, 52*(3), 406–426.

Lagoarde-Segot, T. (2020). Financing the sustainable development goals. *Sustainability, 12,* 2775.

Lee, J., Suh, T., Roy, D., & Baucus, M. (2019). Emerging technology and business model innovation: The case of artificial intelligence. *Journal of Open Innovation: Technology, Market, and Complexity, 5,* 44.

Lindgardt, Z., Reeves, M., Stalk, G., & Deimler, M. (2009). *Business model innovation: When the game gets tough change the game*. The Boston Consulting Group. http://www.bcg.com/documents/file36456.pdf.

Lindič, J., & Marques da Silva, C. (2011). Value proposition as a catalyst for a customer focused innovation. *Management Decision, 49*(10), 1694–1708. https://doi.org/10.1108/00251741111183834.

Louche, C., Idowu, S. O., & Filho, W. L. (2010). *From risk management to value creation*. Greenleaf Publishing.

Lozano, R. (2014). Creativity and organizational learning as means to foster sustainability. *Sustainable Development, 22*(3), 205–216.

Lüdeke-Freund, F. (2010). Towards a conceptual framework of business models for sustainability. In R. Wever, J. Quist, A. Tukker, J. Woudstra, F. Boons, & N. Beute (Eds.), *Proceedings of ERSCP-EMSU conference 2010—Knowledge collaboration and learning for sustainable innovation. European Roundtable on Sustainable Consumption and Production Society*.

Mahajan, R., & Bose, M. (2018). Business sustainability: Exploring the meaning and significance. *IMI Konnect, 7*(2), 8–13.

Marcel, J. (2001). *Sustainable finance and banking: The financial sector and the future of the planet*. Earthscan.

Markides, X. (1999). *Adapted from all the right moves*. Harvard Business School Press.

Mill, G. (2006). The financial performance of a socially responsible investment over time and a possible link with corporate social responsibility. *Journal of Business Ethics, 63*(2), 131–148.

Morris, M., Schindehutte, M., & Allen, J. (2005). The entrepreneur's business model: Toward a unified perspective. *Journal of Business Research, 58,* 726–735.

Mullins, J., & Komisar, R. (2009). *Getting to plan B: Breaking through to a better business model*. Boston, MA: Harvard Business School Press.

Mustafa, R., & Werthner, H. (2011). Business models and business strategy—Phenomenon of explicitnes. *International Journal of Global Business & Competitiveness, 6*(1), 15–30.

Nosratabadi, S., Pinter, G., Mosavi, A., & Semperger, S. (2020). Sustainable banking: Evaluation of the European business models. *Sustainability, 12,* 2314.

Olawumi, T. O., & Chan, D. W. (2018). A scientometric review of global research on sustainability and sustainable development. *Journal of Cleaner Production, 183,* 231–250.

Osterwalder, A. (2004). *The business model ontology—A proposition in a design science approach.* Dissertation 173, University of Lausanne, Switzerland.

Osterwalder, A., Ondrus, J., & Pigneur, Y. (2005, May). *Skype's disruptive potential in the telecom market: A systematic comparison of business models* (HEC LAUSANNE Working paper).

Osterwalder, A., & Pigneur, Y. (2002). An eBusiness model ontology for modeling eBusiness. In *Bled 2002, Proceedings, AIS Electronic Library* (pp. 75–91).

Osterwalder, A., & Pigneur, Y. (2009). *Business model generation.* Amsterdam: Selfpublished.

Osterwalder, A., & Pigneur, Y. (2010). *Business model generation: A handbook for visionaries, game changers, and challengers.* John Wiley & Sons.

Osterwalder, A., & Pigneur, Y. (2012). Designing business models and similar strategic objects: The contribution of IS. *Journal of the Association for Information Systems, 14*(5), Article 3. https://doi.org/10.17705/1jais.00333. Available at https://aisel.aisnet.org/jais/vol14/iss5/3.

Osterwalder, A., & Pigneur, Y. (2013). Designing business models and similar strategic objects: The contribution of IS. *Journal of AIS, 14*(5), 237.

Palo, T., & Tähtinen, J. (2014). A network perspective on business models for emerging technology-based services. *Journal of Business & Industrial Marketing, 26*(5), 377–388.

Porter, M., & Kramer, M. (2011). Creating shared value. *Harvard Business Review, 89*(1/2), 62–77.

Qastharin, A. R. (2015). *Business model canvas for social enterprise.* In The 7th Indonesia International Conference on Innovation, Entrepreneurship, and Small Business (IICIES 2015). https://www.researchgate.net/publication/323393037_Business_Model_Canvas_for_Social_Enterprise.

Ramdani, B., Binsaif, A., & Boukrami, E. (2019). Business model innovation: A review and research agenda. *New England Journal of Entrepreneurship, 2*(2), 79–108.

Roome, N., & Louche, C. (2015). Journeying toward business models for sustainability: A conceptual model found inside the black box of organisational transformation. *Organization and Environment, 29*(1), 11–35. Sage.

Saebi, T., & Foss, N. J. (2014). Business models for open innovation: Matching heterogenous open innovation strategies with business model dimensions. *European Management Journal, 3*(3), 201–213.

Sakis, K., Pinney, C. H., & Serafeim, G. (2016). ESG integration in investment management: Myths and realities. *Journal of Applied Corporate Finance, 28*(2, Spring), 10–16.

Sabatier, V., Mangematin, V., & Rousselle, T. (2010). From recipe to dinner: Business model portfolios in the European biopharmaceutical industry. *Long Range Planning, 43*(2–3), 431–447.

Santos, J., Spector, B., & van der Heyden, L. (2009). *Toward a theory of business model innovation with incumbent firms* (Working paper no. 2009/16/EFE/ST/TOM) (pp. 5–13). INSEAD.

Schaltegger, S., Lüdeke-Freund, F., & Hansen, E. (2012). Business cases for sustainability: The role of business model innovation for corporate sustainability. *International Journal of Innovation & Sustainable Development, 6*, 95–119.

Šimberová, I., & Kita, P. (2020). New business models based on multiple value creation for the customer: A case study in the chemical industry. *Sustainability, 12*(9), 3932. https://doi.org/10.3390/su12093932.

Slávik, Š., & Bednár, R. (2014). Analysis of business models. *Journal of Competitiveness, 6*(4):19–40.

Slywotzky, X. (1995). *Value migration*. Harvard Business School Press.

Steen, M. (2013). Co-design as a process of joint inquiry and imagination. *Design Issues, 29*(2), 16–28.

Stubbs, W., & Cocklin, C. (2008). Conceptualizing a "sustainability business model". *Organization & Environment, 21*(2), 103–127.

Tanwar, R. (2013). Porter's generic competitive strategies. *IOSR Journal of Business and Management, 15*(1), 11–17.

Teece, D. (2010). Business models, business strategy and innovation. *Long Range Planning, 43*, 172–194.

Tikkanen, H., Lamberg, J.-A., Parvinen, P., & Kallunki, J.-H. (2005). Managerial cognition, action and the business model of the firm. *Management Decision, 43*, 789–809. https://doi.org/10.1108/00251740510603565.

Tokarski, A., Tokarski, M., Wójcicki, J. (2017). The possibility of using the business model canvas in the establishment of an operator's business plan. *Torun Business Review, 16*(4), 17–31.

Trimi, S., & Berbegal-Mirabent, J. (2012). Business model innovation in entrepreneurship. *International Entrepreneurship and Management Journal, 8*(4), 449–465.

Uschold, M., & King, M. (1995). *Towards a methodology for building ontologies*. Citeseer.

Voelpel, S. C., Leibold, M., Tekie, E. B., & von Krogh, G. (2005). Escaping the red queen effect in competitive strategy: Sense-testing business models. *European Management Journal, 23*, 37–49. https://doi.org/10.1016/j.emj.2004.12.008.

Watson, D. (2005). *Business models*. Petersfield: Harriman House Ltd.

Wirtz, B. W., Pistoia, A., Ullrich, S., & Göttel, V. (2016). Business models: Origin, development and future research perspectives. *Long Range Planning, 49*(1), 36–54.

World Commission on Environment and Development. (1987). *Our common future* (p. 27). Oxford: Oxford University Press. ISBN 019282080X.

Wu, X., & Sivalogathasan, V. (2013). *Value business model for future: A complex organization to a simple organization*. In 5th International Conference on "Role of Innovation in Business", IFIM Business School in Bangalore, India, 16–17 May 2013. https://www.researchgate.net/publication/265844112_Value_Business_Model_for_Future_A_Complex_Organization_to_a_Simple_Organization.

www. 1. https://www.un.org/sustainabledevelopment/sustainable-development-goals/.

Yip, A. W., & Bocken, N. M. (2018). Sustainable business model archetypes for the banking industry. *Journal of Cleaner Production, 174*, 150–169.

Ziolo, M., Filipiak, B. Z., Bąk, I., & Cheba, K. (2019). How to design more sustainable financial systems: The roles of environmental, social, and governance factors in the decision-making process. *Sustainability, 11*(20), 5604. https://doi.org/10.3390/su11205604.

Zott, C., & Amit, R. (2007). Business model design and the performance of entrepreneurial firms. *Organization Science, 18*, 181–199.

Zott, C., & Amit, R. (2008). The fit between product market strategy and business model: Implications for firm performance. *Strategic Management Journal, 29*(1), 1–26.

Zott, C., & Amit, R. (2010). Business model design: An activity system perspective. *Long Range Planning, Business Models, 43*(2–3), 216–226.

Zott, C., Amit, R., & Massa, L. (2011). The business model: Recent developments and future research. *Journal of Management, 37*, 1019–1042.

CHAPTER 3

Banking Regulations on Financial and Non-financial Risk

This chapter addresses the effects of restrictions of banking regulations on risk on the ability to better integrate banking ESG with corporate ESG goals.[1] This chapter discusses and emphasizes Directives of the European Parliament linked to sustainable finance (among others 2019/2088, 2016/1011, 2019/2089, 2017/828, and 2014/95/UE), The Equator Principles, United Nations Environment Programme Finance Initiative, Collevecchio Declaration, European Union Action plan for sustainable finance, EU taxonomy for sustainable activities, and others related to sustainable reporting, the Principles for Responsible Investment (PRI), solvency, and Basel.

3.1 Introduction

The effectiveness of the implementation of the goals of sustainable development and sustainable finance as well as the implementation of the assumptions of both theories are largely determined by the quality of the legal environment. Legal norms regulating the issues of sustainable development are usually considered on three levels: international law, European Union law, and national law (Barral, 2012; Kenig-Witkowska, 2017). On

[1] The chapter is a part of the project and research results presented in M. Ziolo, Sustainable finance. Development, Risk, Market, PWE, Warsaw 2020 and Financial sphere toward sustainable development. Ed. K. Raczkowski, M. Ziolo, CeDeWu, Warsaw 2017.

© The Author(s), under exclusive license to Springer Nature Switzerland AG 2021

M. Ziolo et al., *Sustainability in Bank and Corporate Business Models*, Palgrave Studies in Impact Finance,
https://doi.org/10.1007/978-3-030-72098-8_3

the basis of international law, a special role is played by resolutions, declarations, and conventions, as well as international jurisprudence focusing on the problems of sustainable development. In the literature on the subject, the view is that legal aspects within the category of sustainable development can be treated as a branch of international law, which results from the dynamically developing jurisprudence in this area and the related analysis of legal acts in terms of their consistency with the postulates of sustainable development.

In the international law environment, the initiatives of the United Nations (UN), including numerous conferences, have played a significant role in creating the legislative basis for the concept of sustainable development (Sustainable Development, 2017). In particular, the UN Conference on the Human Environment (Stockholm Conference) is worth mentioning, which was held in Stockholm in 1972. The effect of the Stockholm conference was the Stockholm Declaration—the resolution of the Stockholm Conference of 14 June 1972 on the natural environment of humans. The Stockholm Declaration is a document that created the foundations for environmental protection, emphasizing the role of education and humans, and, more broadly, the participation of society in shaping attitudes and foundations for environmental protection. The right to a clean environment is in the Declaration, next to the right to property and the right to life, which are equivalent rights. The exercise of rights is determined by the awareness of the public that they have such rights and that they have the necessary legislative means to assert these rights (Brisman, 2011). From the point of view of the content of the Stockholm Declaration, two principles are important from its principles, 1 and 19, which stipulate that it is people who are responsible for protecting and improving the environment (principle 1), and only education can ensure the responsible behavior of individuals, enterprises, and the whole communities in the protection and enhancement of the environment (principle 19). Assigning society responsibility for the condition of the natural environment is important from the viewpoint of the EU subsidiarity principle and the enforcement of its postulates. Pursuant to this principle, it is citizens, who best diagnose and communicate the needs in the sphere of environmental protection and effectively influence the decision-making process, who affect this environment (Sohn, 1973).

Another UN conference on Environment and Development (also known as the Earth Summit), held 3–14 June 1992 in Rio de Janeiro, was also important from the perspective of establishing international law in the field of environmental protection. The measurable effects of this conference (apart from the numerous unfulfilled postulates preceding it) were:

- The Rio Declaration on Environment and Development;
- Global Program of Action, Agenda 21;
- Declaration on the protection of forests;
- Convention on Biological Diversity (legally binding act);
- Convention on Climate Protection (legally binding act).

The Earth Summit introduced concepts important from the point of view of sustainable development, including new social order, cleaner production, and sustainable development. In terms of rank and significance, it is crucial to try to define the concept of sustainable development by defining the principles that constitute this development, including sustainable production and consumption, appropriate demographic policy, new forms of international contacts, and new economic systems. In addition to these three, the Rio Declaration (1992) includes a total of 27 principles of sustainable development. The declaration is not only a set of rules for future development, but also lists obligations in the field of environmental protection. The Declaration emphasizes the inseparability of development and environmental protection, which must be implemented through international agreements protecting the integrity of the global environment and the development system. During the Rio conference, there were also critical voices concerning the overly extensive program of the conference, which did not include the most difficult topics for discussion and the arrangements were not very specific and difficult to enforce. The Greenpeace group also expressed its disappointment and criticism, pointing to the omission in the discussion of the problem of changing the consumption lifestyle in highly developed countries, or the lack of consent to the rapid reduction of CO_2 emissions by the countries in the north.

In 2002, the Rio + 10 summit was held in Johannesburg, the result of which was the Declaration on Sustainable Development, which stated, inter alia, that.

the responsibility for sustainable development rests with the international community;
eradicating poverty, changing consumption patterns, and protecting natural resources is a prerequisite for the implementation of sustainable development; and
as a result of human activity, the condition of the environment is deteriorating on a global scale (e.g., climate change, loss of biodiversity, etc.).

On 20–22 June 2012, the next Rio + Conference, The Future We Want, took place, where the principles set out in the Rio Declaration in 1992 were renewed, and much attention was paid to the institutional dimension of sustainable development. The last important event worth emphasizing is the summit of the 2030 Agenda for Sustainable Development, which occurred on 25–27 September 2015 at the UN headquarters in New York. As a result of this event, the document entitled Transforming our world: the 2030 Agenda for Sustainable Development, which describes, among others, 17 main Sustainable Development Goals. For each of the main goals, operational goals were assigned, of which there are 169 in total.

Agenda 21 is the final document produced by the Rio Conference (1992). It defines the global program of activities for cooperation in development and environmental protection. The document consists of four groups of issues: social and economic issues; issues of protection and management of natural resources; issues concerning strengthening the role of various social groups in the implementation of Agenda 21, including women, youth, trade unions, rural population, local authorities, industry, and science; and issues of sustainable development opportunities. Agenda 21 contains recommendations related to the protection and shaping of the human environment. The document draws attention to the social and economic determinants affecting the quality of this environment and the need to protect natural resources, as well as the need for rational resource management to ensure sustainable and sustainable development. Agenda 21 is an important document as it proposes systemic solutions in terms of general solutions leading to the improvement in the global environment, presenting local problems in relation to the global situation.

The legal approach to sustainable development at the European Community level has evolved. In the first phase, the regulations were harmonized with a view to maintain their consistency due to the leveling of competition conditions in the common market. Only in the second phase was the Environmental Action Program (EAP) strictly in line with the pro-environmental objectives and served the development of European environmental protection standards and provided indicators for assessing the quality of life and the level of pollution.

Important from the strategic and political points of view was the recognition in the Amsterdam Treaty (1997) that achieving sustainable development is one of the objectives of the European Union. Setting this goal at the treaty level significantly changed its rank and positioned it within the essential priorities of the Community. In 2001, during the summit of the European Council in Stockholm, a decision was adopted that sustainable development would become the third element of the Lisbon process, alongside economic and social issues, the aim of which at that time was to position the European Union economy as the most competitive and dynamic world economy, based on knowledge, capable of sustainable economic growth, increasing employment, and improved social justice. These plans were unsuccessful: the crisis of 2008 occurred in the European Union and its individual states drastically deteriorated (in economic, social, and political aspects, etc.), and the Lisbon Strategy was replaced by the document entitled Horizon 2020.

In June 2001 in Gothenburg, an important strategic document was adopted defining the basic threats to the natural environment, referred to as the Sustainable Development Strategy of the European Union. In this strategy, the greatest environmental threats include climate change as a result of greenhouse gas emissions, the presence of chemical compounds hazardous to health in food, poverty and social exclusion, aging of societies, loss of biodiversity, as well as insufficient waste management and pollution related to road traffic. This strategy was implemented through the environmental action programs. The document currently in force is the Strategy for Smart, Sustainable, and Inclusive Growth, Europe 2020. This document considers the perspective of action in the reality due to the 2008 crisis and covers three interrelated priorities: smart growth, sustainable development, and inclusive growth social. The strategy is accompanied by seven flagship projects that support the implementation of the strategic goals, including the following projects: Innovation

Union, Youth on the Move, Digital Agenda for Europe, Resource-efficient Europe, Industrial Policy in an Era of Globalization, An Agenda for New Skills and Jobs, and European Platform Against Poverty.[2]

In the context of environmental protection, the summit of the European Council in Cardiff (1998) is worth mentioning. During the summit, a proposal was made to include environmental protection issues in activities undertaken in all Community policies. The general principles of the EU environmental protection policy are based on the principles of precaution, preventive action, and removal of pollution at the source, as well as on the principle of the polluter pays.[3]

The legal framework for activities in the field of environmental protection is defined by Articles 11 and 191–193 of the Treaty on the Functioning of the European Union (TFEU). Pursuant to the provisions of these articles, the European Union has the right to take action in all areas of environmental policy, in particular with respect to air and water pollution, waste management, and climate change. The scope of EU initiatives in this area is determined by the principle of subsidiarity and the requirement of unanimity in the Council (including budgetary issues, spatial planning of towns and villages, land use, quantitative management of water resources, choice of energy sources, and the structure of energy supplies).[4]

The goal of this chapter is to present the legal framework and the recommendations for the banking sector linked to adjusting banks toward sustainable development and nonfinancial (environmental, social, and governance (ESG) risk management. The remainder of this chapter is organized as follows: in Sect. 3.2, international recommendations supporting the development of sustainable finance[5] are discussed. In Sect. 3.3, European Union law and strategic documents related to sustainable finance are explained and presented. Section 3.4 concludes the chapter.

[2] https://ec.europa.eu/eu2020/pdf/1_PL_ACT_part1_v1.pdf (access 2.11.2020).

[3] https://www.europarl.europa.eu/atyourservice/pl/displayFtu.html?ftuId=FTU_5.4.1.html (access 2.11.2020).

[4] https://www.europarl.europa.eu/atyourservice/pl/displayFtu.html?ftuId=FTU_5.4.1.html (access 2.11.2020).

[5] Sustainable finance means regulations regarding sustainability banks are obliged to follow.

3.2 International Recommendations and Initiatives Fostering Development of Sustainability in Banking

From the point of view of the organizational and regulatory framework for the functioning of sustainable finance, it is necessary to pay attention to voluntary but key initiatives that organize sustainable finance and shape the behavior of financial institutions as entities included in sustainable financial systems. Such initiatives in the international dimension in particular include.

- The Equator Principles (EPs);
- United Nations Environment Program Finance Initiative (UNEPFI);
- United Nations Global Compact (UN GC);
- Principles for Responsible Investment (PRI);
- Collevecchio Declaration (CD);
- Sustainable Reporting (Global Reporting Initiative [GRI]).

The Equator Principles (EPs) are a socially responsible investment initiative by the financial sector, launched in Washington on 4 June 2003 and endorsed by 10 global financial institutions: ABN AMRO Bank NV, Barclays plc, Citi, Credit Lyonnais, Credit Suisse First Boston, HVB Group, Rabobank Group, Royal Bank of Scotland, WestLB AG, and Westpac Banking Corporation (Richardson, 2005).

The EPs provide a framework for voluntary credit risk management and set environmental and social (ESG) risk assessment guidelines for project finance. In another approach, the EPs are principles promoted by the IFC and the World Bank, which aim to provide sustainable development and environmental management practices in projects funded by affiliated financial institutions. Currently, 89 financial institutions from 37 countries have adopted the EPs in their business activities. EPs are subject to change: from 4 June 2013, EPIII was significantly extended compared to EPII (2006). EPIII (2013) includes a total of 10 principles (The Equator Principles, 2013):

- Principle 1—Reviewing and categorizing projects: Projects proposed for financing are subject to internal evaluation considering the criterion of social and environmental impact (social and environmental due diligence). Then, on this basis, they are assigned to three classes of projects: A (high-impact projects), B (medium-impact projects), and C (low-impact projects).
- Principle 2—Environmental and social assessment: For projects classified as group A or B, it is necessary to conduct an additional environmental and social assessment on the basis of which solutions will be proposed to minimize the negative impact of these projects on the environment and society; general recommendations and recommendations on how to reduce the negative impact of the projects on the indicated spheres can be provided both by external companies and independent experts.
- Principle 3—Application of environmental and social standards: At this stage, the project is verified in terms of its compliance with the legal orders, regulations, as well as permits with regard to the social and environmental aspects to which projects in a given country should be subjected. As a result of the verification, it is determined whether the project meets, at least to a minimum extent, the applicability conditions.
- Principle 4—Social and environmental factor management system and EPs action plan: For projects classified as group A or B, it is necessary for the client to design, implement, and implement an environmental and social management system (ESMS). When such a system is adopted and implemented, its compliance with application standards is checked; if the standards are not fully met, the client is obliged to develop and implement an action plan according to the EPs.
- Principle 5—Stakeholder engagement: For group A and B projects, it is necessary to present activities for the effective involvement of stakeholders, including the consultation process in the area of identifying potential project risk and the method of its minimization.
- Principle 6—Complaints mechanism: For group A and B projects, it is necessary to develop a grievance mechanism that could be used by potentially affected entities.
- Principle 7—Independent assessment: For all A-category projects and, depending on the needs, for B-group projects, an independent expert is appointed as a reviewer who assesses documentation and

security systems aimed at reducing the negative impact of the project on the environment.

- Principle 8—Convention/clause: For all groups of projects, clients undertake, pursuant to relevant agreements, to comply with laws, regulations, and environmental and social permits; in particular, to comply with the Equator Principles; and to submit periodic reports on the fulfillment of this criterion by the project;
- Principle 9—Monitoring and reporting: The project's compliance with the Equator Principles is assessed by an independent consultant in the field of environmental and social protection and independent experts.
- Principle 10—Reporting and transparency: This principle is in line with principle 5 and additionally imposes an obligation on the client to publish public information, e.g., on the level of greenhouse gas emissions, etc.

The United Nations Environment Program Finance Initiative (UNEPFI) is another international initiative, an agreement between the United Nations Environment Program and the private financial sector around the world. Signing UNEPFI statements provides the opportunity to introduce environmental issues into the daily operation of financial institutions and emphasizes financing pro-environmental practices and technologies.

The United Nations Environment Program Financing Initiative is a form of partnership between the United Nations (Environment Program) and institutions representing the global financial sector. This initiative was launched in the wake of the Earth Summit, which occurred in 1992 and aimed at promoting sustainable finance. Over 250 financial institutions, including banks, insurance companies, and investors, are working with the UN to better understand today's expectations of the financial sector in terms of sustainable finance. UNEPFI's activities, apart from developing and sharing good practices in the field of financing sustainable development, also focus on coordinating and creating an interface for dialogue between representatives of the financial sector, supervisory authorities, regulators, politicians, and decision-makers to increase the involvement of the financial sector in the implementation of the Sustainable Development Goals, in particular those related to the inhibition of negative consequences resulting from climate change. Since 1994, UNEPFI has been organizing the Global Round Table. It is an event that gathers all entities and institutions dealing with sustainable finance to discuss the most urgent and priority issues in this area.

The formal basis for the UNEPFI initiative is the Declaration of Commitment to which financial institutions join, thus accepting its provisions and the resulting activities. By signing the Declaration, financial institutions openly recognize the role of the financial sector in providing activities for the development of a sustainable economy and undertake to consider environmental and social aspects in all spheres of their activity. In particular, the Declaration refers to three areas of action (UNEPFI, 2016):

1. Sustainable development: The leading role of governments in establishing priorities for action and developing a formal and legal framework for the implementation of designated activities toward sustainable development is indicated. The role of financial institutions is emphasized as those entities that integrate individual sectors of the economy and households through their financial, investment, and trade activities. Attention is also drawn to sustainable development being an institutional obligation and an integral part of corporate governance and good business practices.
2. Sustainable management: Sustainable management refers to activities under all three pillars of sustainable development. In particular, in the case of financial institutions, it consists of supporting and promoting a preventive approach to environmental and social issues, which is expressed in anticipating and preventing the potential negative effects of economic activity on the environment and society. Additionally, as part of sustainable management, financial institutions declare that they comply with all applicable local, national, and international legal regulations relating to environmental and social issues. This simultaneously involves taking steps to integrate environmental and social issues into business activities and decisions in all markets. This means that the identification and assessment of environmental and social threats should be a part of the risk assessment and management process of financial institutions, both in domestic and international operations. Sustainable management also involves the application of best practices in the field of environmental protection and environmental management by financial institutions in their operational activities, including energy and water efficiency; recycling and waste reduction; and establishing business

relationships with customers, partners, suppliers, and subcontractors who adhere to high standards in the field of environmental protection activities. The last element of sustainable management relates to the offer of financial institutions in terms of products and services. In this area, the financial sector sees the need to adjust financial services and products to promote the principles of sustainable development through the product offer.

3. Social (public) awareness and communication: Actions in this regard include recommendations for financial institutions to develop and publish statements on their sustainable development policy and periodically report on the actions they have performed to promote the integration of environmental and social aspects into their activities. Such activities are aimed at providing customers with relevant information to increase their awareness of the possibility of minimizing environmental and social risks and promoting sustainable development.

The UN Global Compact (UNGC) is the world's largest initiative of the United Nations for social responsibility and supporting sustainable development. The UN Global Compact is an initiative supporting and enabling partnerships between governments, civil society, employers, and the United Nations to support and undertake joint actions for sustainable economic development, fight against corruption, protect the environment, and ensure social inclusion.

The Principles for Responsible Investing, or the term Socially Responsible Investing (SRI), are defined by Eurosif as a general term encompassing ethical investments, responsible investments, sustainable investments, and all other forms of investing when investors combine their financial goals with issues related to the environment, society, and corporate governance (ESG). Investing entities aligning with the idea of responsible investing have a written policy on the application of ESG criteria. The principles of responsible investment are accompanied by 35 actions that investors can take to minimize ESG risk in their decisions. These activities address a variety of issues including investment decision-making, asset management, transparency, collaboration, and the achievement of broader support for these practices across the financial services industry. The six core principles of responsible investment for institutional investors are defined as follows (PRI, 2020):

1. incorporating ESG considerations into investment analysis and decision-making processes;
2. being an active owner and integrating ESG issues in property rights management policies and practices;
3. searching for appropriate channels of disclosing information to investors on ESG issues and their impact on the investment process and effects;
4. promoting and accepting the principles of responsible investment in the investment process;
5. cooperation in implementing the principles of responsible investment and their more effective respect;
6. reporting on the implementation of responsible investing principles and the results achieved on their basis.

The Collevecchio Declaration is a declaration for sustainable banking proposed by an informal network of financial institutions. The declaration was adopted in 2003; today, around 200 institutions declare their will to respect its provisions. The declaration constitutes a specific benchmark for financial institutions, enabling them to implement the concept of sustainable banking in the conditions of civil society. The declaration consists of six postulates/obligations that should be respected by financial institutions (Collevecchio Declaration, 2020; Collevecchio, 2003):

- a commitment to activities supporting the concept of sustainability in finance. Financial institutions must extend their missions of action to include environmental and social aspects conducive to ensuring the operation and functioning of a financial institution in accordance with the assumptions of the concept of sustainable finance. Achieving sustainable development and sustainable finance should become goals for financial institutions on par with goals such as ensuring customer satisfaction and maximizing value for stakeholders;
- a do no harm. Financial institutions should undertake actions and decisions and adopt in their activities business models that are consistent with the postulates of sustainable finance and sustainable development and do not violate the environmental, social, and economic order in any way, maintaining a balance between them. This should be expressed, in particular, by the lack of involvement

in dirty sectors and investments, and by entering into cooperation with entities that create negative externalities;

- a commitment to be responsible. Financial institutions should bear the full risk and consequences of their actions and transactions in the context of their impact on the natural and socio-economic environments; this applies in particular to the full assumption by financial institutions of the ESG risk of transactions and the costs of risk creation by this for local and regional communities;
- a commitment to be accountable. The financial institution is accountable to the stakeholders for its decisions and actions for broadly understood sustainability, submitting to legal consequences, and accepting all recommendations and recommendations in terms of changing the style and strategy of operation;
- commitment to transparency. The financial institution should be transparent by providing stakeholders with full access to information about the activities undertaken and their consequences;
- a commitment to good management and ensuring corporate governance favoring sustainability. The financial institution should ensure that it is capable of fostering sustainable development by supporting public policies, regulatory, and/or market mechanisms that ensure the sustainability of complete externality, accounting for social and environmental costs.

The Global Reporting Initiative is a guideline that is a popular worldwide standard for reporting activities related to corporate social responsibility (CSR) and sustainable development. The GRI G3 standard is a set of guidelines specifying the method of reporting the effects of an enterprise's activity, considering three levels (economic, social, and environmental). The standard has two main parts: Reporting Principles and Guidelines and the Sustainability Reporting Guidelines. The first part outlines the recommendations to follow the rules concerning the content and quality of the report, while the second part indicates what information should be presented, with an emphasis on information regarding the organization's strategy and profile, governance structure and management approach, and performance indicators in the structure including economic, social, and environmental indicators.

3.3 European Union Law and Regulations Referring to Sustainability in Banking

Considering the wide range of the impacts of environmental and social risk on the operational activity of banks, regulations supporting the management of these types of risk are necessary. Regulations in the field of sustainable finance adopted by the European Union so far include.

- Regulation (EU) 2019/2089 of the European Parliament and of the Council of 27 November 2019 amending Regulation (EU) 2016/1011 with regard to EU Climate Transition Benchmarks, EU Paris-aligned benchmarks, and sustainability-related disclosures for benchmarks;
- Regulation (EU) 2019/2088 of the European Parliament and of the Council of 27 November 2019 on sustainability-related disclosures in the financial services sector;
- Regulation (EU) 2016/1011 of the European Parliament and of the Council of 8 June 2016 on indices used as benchmarks in financial instruments and financial contracts or to measure the performance of investment funds and amending Directives 2008/48/EC and 2014/17/EU and Regulation (EU) No. 596/2014;
- Directive of the European Parliament and of the Council (EU) 2017/828 of 17 May 2017 amending Directive 2007/36/EC with regard to encouraging shareholders to long-term commitment (Shareholder Rights Directive II [SRD II]);
- Directive 2014/95/EU of the European Parliament and of the Council of 22 October 2014 amending Directive 2013/34/EU regarding disclosure of non-financial and diversity information by certain large entities and groups.

The next legislative steps at the EU level until 2025 concern labels (labeling of financial products), taxonomy, and a uniform standard for green bonds. All of the solutions (binding and planned) directly and indirectly contribute to reducing the risk of non-financial factors in the context of their impact on financial risk. For these activities, the EPs, being non-obligatory, are an important supplement.

Analysis of applicable documents such as Basel II and Basel III (standards for operational risk in banks, monitoring of environmental risk in relation to legal security for repayment), ISO 26000 (reference to the concept of socially responsible [CSR] business), or the GRI and EFFAS indicators for monitoring the effectiveness of ESG risk management confirms that formal grounds exist for further in-depth activities aimed at full integration of ESG risk in the decisions of financial institutions.

In 2018, the European Commission adopted a plan for the financial system to support the EU action for climate and sustainable development. This plan was a response to the need to adjust the financial system to increase the effectiveness of the financing of sustainable development. It is a document corresponding to the entries and postulates of actions for climate protection adopted at the Climate Summit in Paris in 2015 (EC Europe, 2019) and in the 2030 Agenda for Sustainable Development (Agenda for Sustainable Development, 2019). The justification for the plan indicates the increasingly noticeable effects of climate change and their financial consequences. Attention is drawn to the key role of finance in counteracting climate change and in implementing the goals of sustainable development. Apart from the 10 actions set out in the Plan, its three main objectives are underlined as follows (Communication, 2018):

1. reorientation of capital flows and orienting them toward sustainable investment to ensure sustainable and inclusive economic growth;
2. financial risk management created by climate change, resource depletion, environmental degradation, and social issues (including exclusion); and
3. accelerating efforts to ensure transparency and the inclusion of a long perspective of decisions in financial and economic activities.

The commentary to objective 1 raises the problem of the unsatisfactory level of investments ensuring support for the implementation of environmentally and socially Sustainable Development Goals. It is estimated that the financial gap that would have to be covered to achieve the climate and energy goals defined by the European Union by 2030 amounts to EUR 180 billion. The total financial gap covering all other goals, including infrastructure, is estimated at EUR 270 billion. The transformation of the economy toward greening, low-emission, and circular economies requires increasing the amounts of investments supporting this transformation process to approx. EUR 500 billion and allocating funds for financing sustainable infrastructure and innovation in the field of modern solutions supporting climate change (Communication, 2018).

In relation to objective 2, it is assumed that the risk management system needs to be changed so that the system considers the impact and role of non-financial factors in creating transaction risk. Therefore, it is postulated to reorient the decision-making criteria so that, apart from the assessment of economic factors, the importance of environmental and social factors is also considered because their role and importance in the effectiveness of financing, profitability, and asset valuation are systematically increasing (Communication, 2018).

The last of the three objectives discusses the need for the transparent reporting of non-financial factors by all market participants to ensure a more effective selection of potential cooperators with the exclusion of those entities that do not fulfill the requirements of sustainable development in their activity or even conduct socially and environmentally harmful activities. It is also postulated to change the perspective of perceiving and making financial decisions. This change will deviate short-term decisions, actions, and perspectives toward a long-term period of action and perspectives against which decision-makers should address their decisions and assess their effects (Communication, 2018).

The plan of the European Commission is a set of measures used to adapt the financial system to sustainable financing. The 10 listed measures include the following:

1. The need for the European Union to establish a classification system for sustainable activities. Shifting capital flows toward financing sustainable projects requires, first, defining the meaning of the sustainable category. Specifying this concept and precisely explaining what types of activities fit into sustainable projects and initiatives will allow for their classification into homogeneous typological groups. This classification, in turn, will form the basis for funding decisions for specific groups. The main classification criteria will be the pillars of sustainable development, i.e., environmental (with a special role for actions to improve the climate), social, and economic.
2. Developing standards and creating a brand for sustainable banking products. The development of standards and classification of financial products that have been marked as supporting sustainable development will significantly influence the decisions of investors and clients, who will be able to easily decide whether they want to use a balanced or traditional offer of financial institution products.

An example of this is the market of green bonds, which have their brand in the financial market, are recognizable, and the mechanism of their operation is known.

3. Accelerating work on the financing system for sustainable projects. According to OECD data, the infrastructure is responsible for generating 60% of greenhouse gas emissions; therefore, it is important to develop a basis for financing sustainable investments that will allow reducing the emission value by building a sustainable and environmentally friendly infrastructure. To support the financing of sustainable investments and the accumulation of private funds, the European Investment Advisory Hub was established.
4. Incorporation of sustainability issues in the financial advisory process. The role of financial advisers and intermediaries in disseminating the concept of sustainable development is crucial. These entities significantly influence the financial choices and decisions of clients of the banking and financial sector in the field of financial products and services. Advice on structuring transactions and selecting sources of financing or investor investment strategies contributes to more conscious operation and adaptation of the financial market to the needs of sustainable development.
5. Developing the sustainability benchmarks. Balanced benchmarks are important from the viewpoint of the valuation of financial instruments and information for investors who use them to make decisions and follow the direction of changes in the financial market. The traditional benchmarks have been adjusted for the purposes of assessing the impact of non-financial factors (ESG) by including these factors in the methodology of benchmark development. Due to the lack of transparency in the methodologies adopted in these activities, sustainable benchmarks must be systematically improved and developed.
6. Better integration of sustainability concerns in ratings and market research. The financial market saw the need to verify the current approach to developing ratings, testing creditworthiness, and conducting market research, consisting of considering non-financial factors (ESG) in the methodologies, procedures, and analyses. Especially in their assessment methodologies, sustainable rating agencies analyze and consider the impact of environmental, social, and management factors, and track their evolution over the years. Banks, in turn, in their internal ratings prepared for the

purposes of assessing creditworthiness, also verify the impact of ESG factors on the financial situation of clients and on the value of assets, especially those that are to be the subject of transaction security. The activities of rating agencies are monitored by the European Commission in terms of the adopted solutions and methodologies on the basis of which the ratings and rating outlook are formulated.

7. Identification of liabilities of institutional investors and asset managers. Institutional investors and asset managers under the laws of the European Union are committed to acting in the best interests of clients and investors. However, the issue of regulation is unclear in terms of non-financial factors and the associated risks. In this context, the responsibilities of institutional investors and managers toward their clients are not clearly defined. As practice shows, not all managers and investors take action and include ESG factors in the process of their decisions.
8. Integrating sustainability into the prudential requirements of banks and insurance companies. Banks, insurance companies, and pension funds belong to the group of institutions that create and provide the main transfer channels for savings and investments. These institutions have the potential to bridge the investment gap in financing sustainable infrastructure. These entities may simultaneously be strongly exposed to risk from non-financial factors, in particular those responsible for climate change. This risk has its dimension and financial consequences: it may affect the financial situation of the indicated entities and thus affect financial stability and macroprudential risk. Therefore, all regulations in the field of macro- and microprudential supervision, as well as procedural actions aimed at ensuring financial stability, should be directed at minimizing this type of risk.
9. Paying more attention to disclosure of non-financial information. Non-financial reporting is a key source of information on entities and activities in the field of supporting the implementation of Sustainable Development Goals, as well as those activities that are harmful and have a negative impact on the environmental and social spheres (including negative externalities). Therefore, it is necessary to develop a legal framework and standards of operation based on which it will be possible to collect comprehensive

and high-quality information on nonfinancial factors in the activities of market entities. Such information will provide an important signal for market participants and, more broadly, stakeholders who, guided by the premises of sustainable development, will make market decisions.

10. Accelerating the process of sustainable corporate governance and limiting short-term decisions on capital markets. Corporate governance plays an important role in ensuring the development of sustainable economy and sustainable growth, it enables enterprises and other market entities to adjust business models in terms of creating sustainable value and increasing operational efficiency. This adaptation approach allows for more-adequate risk management and increasing competitiveness. However, the challenge for enterprises is extending the perspective to making decisions. In this context, the financial market, by offering financial instruments dedicated to long-term activities, may motivate entrepreneurs to plan using a long-term operating horizon.

Following specific measures to ensure the financing of sustainable growth, legislative work on the development of a legal framework for the sustainable finance system was initiated. The first legislative proposals concern.

- disclosure of information on sustainable investments and risks for sustainable development and on the amendment of Directive (EU) 2016/2341;
- low-carbon benchmarks and positive carbon impact benchmarks, and amendment of Regulation (EU) 2016/1011;
- establishing a framework to facilitate sustainable investment.

The first of the priority actions on the list of activities, i.e., the development of a classification system for sustainable economic activities, was completed in December 2019. Taxonomy is the first document that systematizes economic activities in terms of the degree of their impact on environmental goals and the sustainable economy. The taxonomy regulation sets out the framework to ensure the development of an EU classification system for environmentally sustainable economic activities. The economic activities are qualified in accordance with the following environmental objectives:

- mitigating climate change;
- adaptation to climate change;
- sustainable use and protection of water and marine resources;
- transition to a circular economy;
- pollution prevention and control;
- protection and restoration of biodiversity and ecosystems.

An economic activity can be considered sustainable when it meets at least four criteria (EC Europe, 2019, p. 19):

- provides a significant contribution to the achievement of at least one of the six defined environmental objectives,
- does not significantly damage other environmental objectives,
- complies with the technical criteria, and
- compatible with minimum guarantees related to social security and management.

The taxonomy is intended to contribute to ensuring Europe's status as the first climate-neutral continent by 2050 and will be obligatory for.

- Member States and the European Union;
- financial market participants offering financial products;
- financial and non-financial companies covered by non-financial reporting.

The Taxonomy Regulation affects financial market participants by obliging them to report how they affect sustainable economic activities. The financial market participants include.

- insurance companies that provide an insurance-based investment product (IBIP),
- alternative investment fund managers (AIFM),
- investment companies offering portfolio management,
- IORPs or pension product providers,
- managers of a qualifying venture capital fund,
- managers of a qualifying social entrepreneurship fund, and
- UCITS management companies.

The Taxonomy Regulation imposes obligations related to equity and debt investment vehicles offered as environmentally sustainable or with similar characteristics. Investments in private equity, real estate funds, and private securitized loans may also be subject to regulation if the funds obtained are sold as green. The taxonomy may also be used optionally by other financial entities, such as banks, for project finance or debt issuers (or green income, capital expenditure [CAPEX], sustainable assets, or activities reporting entities). The summary of the regulations and initiatives discussed in this chapter presents Table 3.1.

3.4 Conclusion

It should be emphasized and positively assessed that initiatives and actions standardize, unify, and offer ready-made solutions for financial institutions in the field of ESG risk management. This means that financial institutions are more protected and reduce their exposure to this type of risk and have a basis for developing good practices. However, there is the problem of the costs of implementing the proposed solutions, their maintenance and monitoring, and, importantly, the problem of subsequent regulations of the financial sector, which, especially after the 2008 crisis, is heavily regulated. The proposed solutions mainly concern large financial institutions, the scale and scope of which are justified by being exposed to risk, including ESG risk. Smaller, local financial institutions will face the dilemma of financing the proposed solutions and the lack of resources to implement them. ESG risk management standards or the taxonomy of sustainable economic activities should be assessed positively due to the homogeneity of the approach that will ensure the comparability of the effects of initiatives and actions taken by individual EU countries. This will become the basis for obtaining data that will allow constructing a comprehensive picture of the scale of the problem to be solved and the need to manage ESG risk, mainly environmental risk, in the areas and sectors identified as problematic. At present, such information is lacking, which makes many of the actions taken intuitive. Simultaneously, the taxonomy and procedures of ESG risk management in financial institutions are expected to contribute to revealing the problem of environmentally sensitive sectors, which will face the challenge of adjusting business models toward more sustainable ones. Another important issue is the effectiveness of the impact of individual initiatives and activities that have different ranking (optional vs. obligatory). In the case of optional

Table 3.1 Scopes of regulations on financial and non-financial risk

Scope of regulations	*Type of document C-compulsory F-facultative*	*The purpose*	*Impact on*	*Impact on greening economy D—direct I—indirect*
Climate transition	C	Transition benchmarks	Banks	D
Sustainability-related disclosures in the financial services sector	C	Sustainability-related disclosures in the financial statements	Banks	D
Benchmarks in financial instruments and financial contracts	C	Measurement the performance of investment funds	Investment funds	I
Long-term shareholder engagement	C	Encouragement of long-term shareholder engagement	Companies, including stock companies	I
Disclosure of non-financial and diversity information	C	Disclosure of non-financial and diversity information by certain large undertakings and groups	Business, banks	I
Environmental risk	C	Environmental risk management	Banks	D
Social risk	F	ISO 26000 is defined as the international standard developed to help organizations effectively assess and address social responsibilities[a]	Business, banks	D
ESG risk	F	The Equator Principles (EPs) is a risk management framework, adopted by financial institutions, for determining, assessing and managing ESG risk[b]	Banks	D

(continued)

Table 3.1 (continued)

Scope of regulations	*Type of document C-compulsory F-facultative*	*The purpose*	*Impact on*	*Impact on greening economy D—direct I—indirect*
Sustainable finance, sustainable financial systems	F	UNEPFI contributes the perspectives of financial institutions to the various United Nations and global activities on sustainable finance[c]	Banks	D
Financial institutions and sustainability	F	The Collevecchio Declaration calls on banks to measure the environmental and social impacts of their decisions, take full responsibility for the environmental and social impacts of their transactions[d]	Banks	D
Sustainable activities	C (draft in progress)	The EU taxonomy is a classification system, establishing a list of environmentally sustainable economic activities	Business	D

[a]https://asq.org/quality-resources/iso-26000 (access 31.12.2020)
[b]https://equator-principles.com/about (access 31.12.2020)
[c]https://www.unep.org/explore-topics/green-economy/what-we-do/finance-initiative (access 31.12.2020)
[d]Collevecchio Declaration—BankTrack www.banktrack.org (access 31.12.2020)

solutions, with their provisions being largely general, it is expected that the effectiveness of their impact in the context of the results obtained will be unsatisfactory in relation to the expectations declared during their establishment (including the EPs).

References

Agenda for Sustainable Development. (2019). https://sustainabledevelopment.un.org/sdgsummit. Accessed 31 Dec 2020.

Barral, V. (2012, May). Sustainable development in international law: Nature and operation of an evolutive legal norm. *European Journal of International Law, 23*(2), 377–400. https://doi.org/10.1093/ejil/chs016.

Brisman, A. (2011). Stockholm conference, 1972. In D. K. Chatterjee (Eds.), *Encyclopedia of global justice*. Dordrecht: Springer. https://doi.org/10.1007/978-1-4020-9160-5_655.

Collevecchio. (2003). *Implementing the Collevecchio declaration on financial institutions and sustainability, 2003.* https://www.publiceye.ch/de/news/detail/implementing-the-collevecchio-declaration-on-financial-institutions-and-sustainability. Accessed 2 Nov 2020.

Collevecchio. (2020). *Collevecchio declaration, the role and responsibility of financial institutions*. BankTrack. https://www.banktrack.org/download/col levechio_declaration/030401_collevecchio_declaration_with_signatories.pdf. Accessed 2 Nov 2020.

Communication from the Commission to the European Parliament, the European Council, the Council, the European Central Bank, the European Economic and Social Committee and the Committee of the Regions. Action Plan: Financing Sustainable Growth. European Commission, Brussels, 8.3.2018 COM. (2018). 97 final.

EC Europe. (2019). *Paris climate conference (COP21).* https://ec.europa.eu/clima/policies/international/negotiations/paris_en. Accessed 31 Dec 2020.

ES Europe 2019, EC Europa. (2019). *EU sustainable finance taxonomy—Technical report*. https://ec.europa.eu/info/sites/info/files/business_economy_e uro/banking_and_finance/documents/190618-sustainable-finance-teg-rep ort-taxonomy_en.pdf. Accessed 2 Nov 2020.

Kenig-Witkowska, M. (2017). The concept of sustainable development in the European Union policy and law. *Journal of Comparative Urban Law and Policy, 1*(1). Study Space IX Cape Town, South Africa Article 6, pp. 64–80.

PRI. (2020). *The un principles for responsible investment and the OECD guidelines for multinational enterprises: Complementarities and distinctive contributions*. https://www.oecd.org/investment/mne/38783873.pdf. Accessed 2 Nov 2020.

Richardson, B. J. (2005). Equator principles: The voluntary approach to environmentally sustainable finance. *The European Environmental Law Review, 14,* 280.

Sohn, L. B. (1973). The Stockholm declaration on the human environment. *The Harvard International Law Journal, 14*(3), 423–514.

Sustainable Development. (2017). https://academic.oup.com/ejil/article/23/2/377/487236/Sustainable-Development-in-International-Law. Accessed 2 Nov 2020.

The Equator Principles. (2013). *The equator principles June 2013, A financial industry benchmark for determining, assessing and managing environmental and social risk in projects*. www.equator-principles.com. Accessed 2 Nov 2020.
UNEPFI. (2016). Connecting financial system and sustainable development: Market leadership paper (p. 3). https://www.unepfi.org/wordpress/wp-content/uploads/2016/11/MKT-LEADERSHIP-REPORT-AW-WEB.pdf. Accessed 2 Nov 2020.

CHAPTER 4

Corporate Social Responsibility and Sustainable Value Creation

The main purpose of this study is presentation of the different perspectives on the implementation of the CSR concept. It refers to enterprises and supply chains. Different perspectives of defining the concept and different strategies and principles were indicated. Various categories of barriers that prevent or hinder the implementation of CSR in organizations were also indicated. Considerable attention has been given to identifying and describing trends in the development of CSR strategies, including in particular the relevance of CSR 2.0 aspects for contemporary organizations. The last part of the chapter focuses on creating sustainable value, its identification and foundations. The presented tools for assessing and identifying sustainable value are to become a helpful catalog for identifying and assessing the level of this particular value in enterprises. Systematizing available scientific knowledge will be helpful material for both understanding and implementing CSR concepts in organizations. The chapter was written using the available literature on the subject, on the basis of which it drew its own conclusions and findings.

© The Author(s), under exclusive license to Springer Nature Switzerland AG 2021

M. Ziolo et al., *Sustainability in Bank and Corporate Business Models*, Palgrave Studies in Impact Finance,
https://doi.org/10.1007/978-3-030-72098-8_4

4.1 Introduction

CSR is a strategy implemented by companies and supply chains, and its perspectives differ depending on the point of view of the different stakeholders and the different implementation paths of individual stakeholders. CSR strategy is not only about benefits, but also about costs, risks, and specific barriers that must be overcome in its implementation. The above elements have different meanings and burdens for different stakeholders, so it seems important to identify them and indicate how they can contribute to the creation of added value, with a particular focus on sustainable value in innovative business models. This chapter identifies the elements of benefits, costs, and risks of the above strategy to the different beneficiaries, as well as identifies the development trends and their impacts on the creation of added value of organizations and their supply chains. The chapter defines the sustainable value produced by social and environmental aspects for organizations, as well as present key drivers for implementing CSR in innovative business models.

4.2 CSR Strategy: A Varied Perspective Point of View

CSR is becoming the important part of organizational strategy that is committed to the benefit of stakeholders through various activities (Goyal & Kumar, 2017). In the literature, CSR (in classical terms) is often described as a pyramid of CSR (Carroll, 1991), in which CSR practices can be divided into four categories: economic, legal, ethical, and discretionary responsibility (Fig. 4.1).

CSR principles and assumptions are relevant to every type of organization, including profit, non-profit, and non-governmental organizations (NGOs). The literature indicates that CSR is the for-profit organizations' most innovative part of any free society's economy (Chandler, 2019).

Over many years of introducing and developing the principles and basic assumptions of CSR, the view on CSR has changed, which in turn has led to a redefinition of the concept and the definition scope. In the current sense, a relationship exists between corporations (both for-profit and non-profit) and societies, which must interact with other elements and responsibilities. The social aspect is important because it covers all stakeholder groups, from customers and employees, to suppliers, creditors, media, and local communities. Within these areas, a trade-off relationship

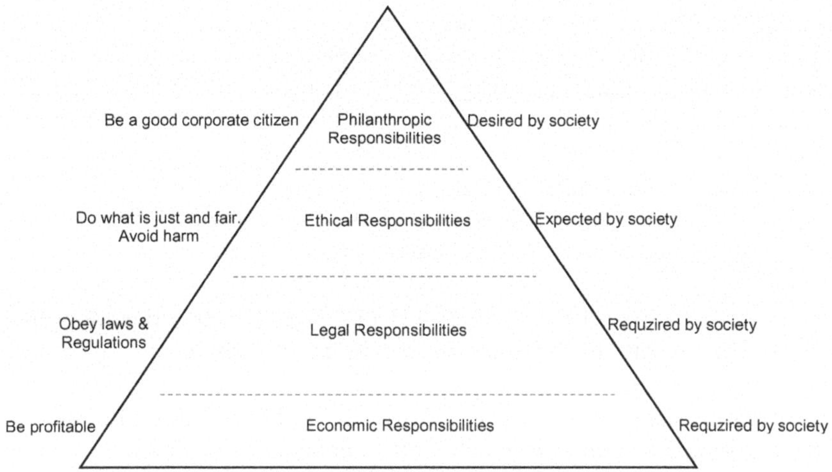

Fig. 4.1 Caroll's CSR pyramid (*Source* Carroll [2016])

must exist for companies so that they can benefit and create value through CSR as a consequence of the new strategy (Chandler, 2019). What is now important is that the scope of CSR should not be based only on those required by legal regulations, but also on the common social good (McWilliams & Siegel, 2000). The identification of stakeholders, their environment, and the requirements and monitoring of undertaken activities facilitate the implementation of CSR principles and, in the long-term, allows for the identification and reduction of barriers, survival, expected growth of activities, and corporations' prosperity. Currently, the definitions and assumptions regarding CSR are convergent regardless of the point of view of the organization creating it. Their goal is the common good and to act on behalf of all stakeholders.

CSR can be considered in different dimensions. The definition of CSR is based on the aspect of the evolution of the concept and its redefinition and on the extension of its scope. There is no boundary framework (area or definition) within which the CSR concept can be contained. Most often in the literature, CSR refers to the following dimensions: stakeholders, social, economic, voluntariness, and environmental (Dahlsrud, 2008); economic, ethical, social, stakeholders, sustainability, and discretionary considerations are the core dimensions of the CSR definitions (Sarkar & Searcy, 2016). The approach to understanding (including defining) CSR can be based on various dimensions. An interesting approach is the ontological approach, containing five categories, within which all aspects should be considered (Yadlapalli et al., 2020). The five categories include three lower-level categories:

(1) the approach is similar to the types/principles of CSR,
(2) the benefits are equivalent to the consequences of CSR, and
(3) considering stakeholder as beneficiaries.

The next two categories are additional higher-level categories identified to form the meaningful CSR communication and include:

(4) temporality, the experience of time in implementing CSR; and
(5) action, the degree to which CSR benefits the stakeholders.

The components of each category are presented in Table 4.1.

This approach allows creating multidimensional definitions of CSR covering very different areas and elements. Considering these areas and the individual elements describing them perfectly reflects the essence of CSR while allowing creating individual definitions in this area. This approach indicates that restriction to a generalized, strict, and closed CSR definition is not possible. The methods of creating definitions using the above areas are presented in Fig. 4.2.

Combining individual elements, categories, and levels of CSR allows for a broad understanding and the application of CSR principles in organizations. All business decisions have both economic and social consequences. It is important to take on the challenge and tasks of managing the emerging conflict of interest between stakeholders and ever-changing needs to achieve synergistic effects. In social terms, it is necessary to find a balance between economic and social and environmental aspects. Such opportunities are provided by CSR, which is constantly evolving in terms of the validity of its principles and its comprehensiveness.

When referring to CSR in global terms, different societies define the relationship between business and society differently. Depending on the economy, various factors play different roles in CSR. Poorer economies will focus on the lower-order needs; in terms of meeting CSR challenges, and richer economies will pay attention to the higher-order goods and while meeting the needs of the higher Maslow hierarchy. Different

Table 4.1 CSR dimension and sub-dimension of different category—ontology approach

CSR dimension and sub-dimension			
Categories	*Dimension*	*Sub-dimension*	
Approach	Altruism	Discretionary; Voluntary; Philanthropic	
	Ethical	Moral mandatory; Desirable	
	Legal	Compliance; Obligation	
	Strategic	Economic; Social audits	
Benefit	Economic	Increased revenues, Cost savings, Productivity, Profit	
	Social	Society, Customer attraction and retention; Fair operating practices; Human resources; Discrimination	
	Environmental	Efficient resource use; Waste reduction; Product life cycle management; Risk reduction	
	Goodwill		
Stakeholders	Primary	Shareholder/Investor; Employee; Customer; Supplier	
	Secondary	Government; Competitor; Consumer advocate; Environmentalist; Special interest group; Media; Community	
Two additional higher-level categories			
Categories	*Dimension*	*Categories*	*Dimension*
Temporality	Episodic; periodic; continuous; sustainable	Action	Contribute; improve; balance; generale; commit; communicate; impact

Source Own elaboration on the basis of: Yadlapalli et al. (2020)

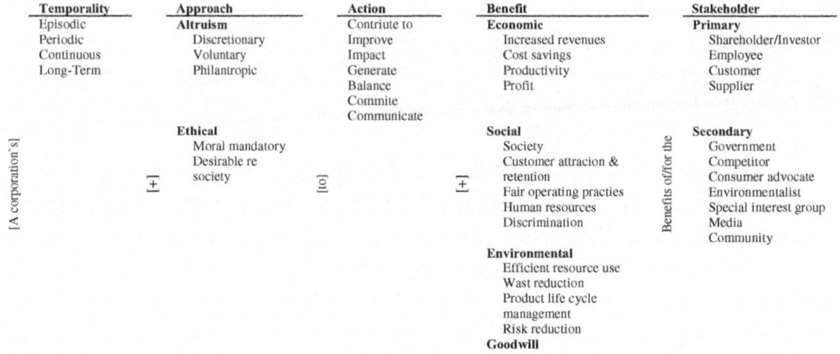

Fig. 4.2 Multidimensional definitions of CSR (*Source* Yadlapalli et al. [2020])

expectations regarding the scope of CSR and areas of application lead to social aspects having different priorities in different countries as well as in different companies or supply chains. It is in the best interest of the organization to adapt to and meet the changing expectations of customers. Often, it is the expectations that create conflict related to different areas of CSR. Therefore, CSR can be considered from the point of view of implementation and management of the arguments raised by organizations: ethical, moral, rational, and economic (Chandler, 2019).

The strategy, principles, and individual elements of CSR can be considered from different perspectives. Most often in the literature, authors refer to the perspective of stakeholders and the tasks that are assigned to them within the CSR strategy. Social responsibility initiatives are dedicated to solutions, to protect consumer rights, or to protect the environment. Nowadays, CSR prospects not only support protection but also focus on the social aspect: this is one of the basic objectives of organizations or their supply chains, which must treat the above assumptions and objectives on par with economic or environmental objectives. This approach indicates the importance of sustainable development and the implementation of particular elements in creating such value in the strategy of the organization, which, among others, bases its activity on clear and specified CSR goals.

The perspectives for the application of CSR include not only the soft elements written in the organizational or chain strategy, but also increasingly focus on the quantitative (hard) measurement of the fulfillment of individual goals. Thus, measures to ensure supplier obligations in supply chains are examined. These analyses not only apply the management perspective, but also network analysis or the concept of integrated environmental decision-making. Demonstrating strength and achieving CSR objectives also include the use of mathematical modeling and programming (including links to CSR performance and, for example, profit levels).

The prospects for implementing CSR may apply to different structures in the organization and its chains. CSR is implemented from initiative at the management level to support the analysis of how the relationships between different stakeholders (internal and external) affect the motivation of enterprises to undertake CSR activities. Control activities and incentives from leaders support, e.g., suppliers in achieving CSR objectives, as well as more effective and less costly activities. The principles can relate to and support quality management of products provided by suppliers. Another perspective for implementing CSR is responsible delivery, which can affect competitiveness. Another aspect influencing the implementation of CSR principles is focused on environmental technology selection and innovation, e.g., in terms of the price and technological choices of the producer for a product that is suitable for reprocessing, or the implementation of technological innovations in the context of environmental policy for reducing emissions or operating under conditions of uncertainty. Current trends in the development of the organization are focused on incentives to transition from traditional to sustainable and CSR-centered activities (Bian et al., 2020). In a sense, the concept of sustainability and its requirements, which are now being implemented in organizations, have contributed to placing pressure on businesses operating in the various economies (Villena, 2019), but the effects of this pressure include sustainable value and new business models that support CSR and sustainability. Regardless of the definition of the concept, the area of corporate social responsibility (CSR) is a crucial issue in business operations (Van Wassenhove, 2019).

In the current economic situation, an organization's attitude toward CSR is largely dependent on the relationship between social responsibility and profit. If CSR practice and its principles and actions contribute to the growth of profits and prosperity, organizations will have a positive attitude toward CSR and will more willingly and to a greater extent implement its principles. Importantly, profits for one organization do not have to be achieved at the expense of other stakeholders. A firm adopting CSR practices can develop a positive brand reputation and can hopefully generate better financial performance (Chan et al., 2020).

Stakeholders' expectations determine the practices to be used in the field of CSR. CSR strategies can be considered from various aspects. The most commonly used ones include the following (Zhang et al., 2019):

- Shareholder interests, including shareholder legal revenues, sustainable development of corporate, decision-making participation, and the shareholder relationship management system;
- Environment preservation, including conservation of energy and resources and emission reduction, biodiversity and ecology protection, environmental management, and green design;
- Well-being of the local community, including project impact on the community and community involvement;

- Employee interests, including the occupational health and safety of employees, wages and welfare, staff employment, education and training, freedom of association and bargaining, and caring for all employees;
- Customer interests, including quality and safety of construction product, and customer service and satisfaction;
- Supplier or partner interests, including enhancing communication and collaboration with suppliers or partners and supply chain management;
- Government commitment, including compliance management, providing employment opportunities, technology and innovation and industry development, and engaging in urban development programs;
- Fair operation and competition, including ethical operation, risk management, fair competition, and protection of intellectual property rights;
- NGO practices, including social and public service strategy and media disclosure; and
- CSR institutional arrangement, including CSR institutional organization, CSR research, and CSR communication.

An important element in considering the CSR concept is the definition of driving forces. Driving forces contribute to CSR activities even if companies do not want to take such initiatives under other conditions. The driving forces for implementing a CSR concept can be divided into the following (Zhang et al., 2017):

- Market drivers: the attitudes of its consumers, employees, investors, business suppliers, and customers;
- Social drivers: pressure from NGOs, media attention, general social expectations, and business coalitions, as more collective social drivers worth considering separately;
- Governmental drivers: traditionally defined as voluntary actions that companies take to contribute to a better society and a cleaner environment; in recent years, many governments have taken policy initiatives to encourage and promote CSR;

- Globalization: in particular, the role of multinational enterprises in providing a push toward standardization of CSR practices across the world;
- Institutional: CSR depends at least partly upon the institutional context of a country. We have shown above how governments, civil society organizations, international organizations, etc., drive CSR. Religion (an informal institution) has long influenced the role that businesses play in society.

In terms of globalization, some of this standardization has arisen through multinational enterprises (MNEs) and foreign direct investment (FDI), increasing trade and the attendant need to meet international regulations if goods/services are to be traded, and some have arisen through the role of international organizations like the UN, World Bank, OECD, the World Trade Organization (WTO), and so on. Globalization might encourage CSR by spreading the business attitudes from developed to developing countries because businesses that trade in foreign countries need to build reputations as good citizens and because there are more international watchdogs (NGOs, ethical investment firms, and even the World Bank, etc.) standardizing these rules across countries. This is more efficient in given varied expectations.

Countries are often classified into rule- and relationship-based governance environments by institutional theory (Dixit, 2009). Rule-based governance environments are those where formal institutions are strong; in these environments, monitoring CSR activities is likely to be effective. In relationship-based environments, informal institutions (like community bonds, trust, and religion) are stronger. In these environments, CSR is more likely to manifest in entirely voluntary philanthropic or community-focused activities. It is less likely to be embedded in a new way of operating. In most countries, CSR began as philanthropy (at a time when societies were relationship-based) and developed into a whole new way of doing business (in rule-based societies).

Summarizing the reflections on the driving forces of CSR, the most important actions include policy pressure, market pressure, innovation and technology development, government legislation, pressure on employers to engage in CSR, the potential for competitive advantage, customer expectations, and long-term investment. Drivers can also be considered in terms of their subjectivity and objectivity. The subjective factors include integrating ethics into a firm's management, integrating transparency and affecting investor confidence, and contemplating aspects related to social, environmental, and economic performance. The objective factors are sectorial trends, improvement in the firm's reputation, social demand for increased transparency, honesty, ethics, publication of sustainability reports by leading corporations, and associated legal and institutional frameworks (Agudo-Valiente et al., 2017).

The description and definitions of CSR, as well as its practical approach, allow for distinguishing individual benefits of implementing this strategy. The most important benefits include better brand recognition, positive business reputation, increased sales and customer loyalty, operational cost savings, better financial results, greater ability to attract talent and retain employees, organizational development, and easier access to capital. Additional benefits include product innovation, employee engagement, increased productivity, employee retention, increased sales, consumer preference, gaining new business, enhanced relationship with stakeholders, media interest, good reputation, access to funding opportunities, enhancing influence in the industry, differentiating yourself from the competitor, saving money on energy and operating costs, increased customer retention, increased visibility, local communication, consumer links and interaction, and strong overlaying strategy. In the context of benefits, we can talk about their division into two categories, for society and for enterprises (Table 4.2).

Other elements related to the implementation of CSR include motivators. Motivators are factors closely related to the main objectives of the organization. Some of them relate to profit maximization, others to activities related to sustainable development. Their main task is to encourage organizations to implement practical CSR solutions. They allow for the implementation of CSR even if so far there was previously a lack of such activities in the organization. The motivating factors are able to encourage the use of new, innovative practices, even without exerting pressure from outside or from other stakeholders. They are closely related to the benefits perceived as a direct effect of CSR implementation. The most frequent motivators for undertaking CSR activities and implementing its elements include strategic business direction, financial benefits, branding, reputation, image, human resource benefits, supplier-induced benefits, persuasion and inspiration, relationship-building, partnership, policy benefits, organizational culture, and awareness.

An important element is the identification of barriers that may contribute to the failure of CSR activities or constitute a reason for not

Table 4.2 CSR benefits for companies and society

CSR benefits	
Company	*Society*
Building a brand image Making profits Licence to operate Avoiding government regulations Employee motivation Differentiation from other brands Attracting investors Cost reduction Good relations with stakeholders Improved quality Customer loyalty Desire to "do good" Risk mitigation Avoiding legal actions	Technology and infrastructure Education Economic development Sense of security Increased health Better employment opportunities Social activation Higher standard of living Improved quality Natural environment protection

Source Own elaboration on the basis on: Księżak (2016)

taking any action in this area. Many categories of different barriers are expressed in the literature. The most common include government policy, stakeholder perspective and expectations, attributes of CSR, lack of top management commitment, lack of knowledge, insufficient CSR training, lack of resources, lack of CSR strategy, lack of organizational structure, lack of concern for reputation, uncertainty about effectiveness of CSR implementation, the complexity involved in CSR implementation, lack of regulations and standards, the high cost of CSR implementation, the time required for implementation, customers not caring about CSR, lack of policy incentives, insufficient human resources (HR) for CSR, lack of stakeholder awareness, lack of CSR consultants, unstable political climate, and resistance to changes (Bux et al., 2020).

From a different perspective, the barriers can be applied in subjective and objective aspects (Agudo-Valiente et al., 2017). The subjective aspects of CSR are especially applicable in large corporations: only focusing on undertaking social- and solidarity-based projects, which is an image strategy rather than a true strategic conviction; considering CSR only a passing fashion that arose after the latest scandals or as a utopian value that cannot be integrated into or managed by firms.

The cost–benefit ratio is a barrier to implementing the CSR concept. This barrier arises when companies do not have the financial resources to develop CSR or when they have to make compromises between profits and actions, e.g., ethical ones (Laudal, 2011). A barrier can also be created when an enterprise does not have adequate knowledge of CSR integration (external control), and when there is a lack of strong leadership and management to integrate CSR principles (internal control).

Looking at the reasons for the barriers, it can be pointed out that in terms of (Goyal & Kumar, 2017):

- Lack of financial resources: CSR investments do not maximize the company's profits; they are only cost-related and often considered as expenses, and the lack of cost efficiency associated with CSR creates a barrier to it;
- Lack of top management commitment: Decisions related to the implementation of CSR are strategic, so they must be supported and implemented by top management. The top management not only initiates and implements concepts, but is also key to motivating other employees. Thus, a lack of support is a barrier;
- Lack of knowledge about CSR practices: This is related to the lack of unambiguity in the definition, the lack of indication of specific assigned categories of actions, the subjective characteristic of actions taken, and the lack of reference to practice;
- Lack of effective strategic planning for CSR may have serious consequences in terms of achieving future competitive advantage and may cause problems in the implementation of individual CSR objectives;
- Absence of significant benefits for CSR implantation: So far, the relationship between CSR implementation and corporate performance has not been proven. This situation may cause dilemmas. There are findings showing that CSR does not benefits the company in the short-term, quantitative measures are lacking, and reports are rather based on qualitative aspects, which discourages the implementation of the concept;
- Complexity of CSR implementation: The concept should be treated as a comprehensive approach, and the potential benefits may depend on the sector, type of activity, size of business units, or problems faced by the national economies concerned, as well as environmental or social problems in the business practices concerned;

- Time-consuming CSR implementation: A long time is needed to implement the concept, the planned changes must be accepted by all levels of the organization, and in all areas, there must be information about the planned changes. The actions taken also differ in nature; often, the actions taken require additional information so that stakeholders understand their meaning and long-term benefits;
- Lack of training and employee participation in effective CSR implementation: The customer's perception of CSR may be complex and multifaceted, which may be an element that determines the choice of a given product and distinguishes it from other products offered by competitors. This may or may not be the case, which is why customers need to be convinced of the importance of the CSR concept. Often consumers know little about the initiatives of a given organization;
- Lack of necessary skill/education: Implementing CSR principles is a complex and difficult task requiring specific skills. Implementation should be based on the knowledge of experts to meet the requirements and properly implement the principles. In this respect, the firm should invest in training, as lack of knowledge may lead to failure to implement CSR.

An approach to overcoming these barriers is their division into those of a psychological nature (Reynolds, 2019): cognitive and emotional. In the former type, cognitive barriers, it is necessary to understand the essence of the company's activity and its goals, which are not only financial but also social, providing a dual and multidimensional view of reality. In this approach, neither the goals nor the scope of activities need to be linked, yet both aspects are necessary for the organization to operate. Another element of CSR is struggling with the abstractive quality of values, which is a factor that does not fully allow managers to follow all the set rules related to the ambiguity of values. CSR requires making difficult choices and is shaped by many positions and elements, each of which has its own character. The choices of values can include abstract concepts (life, freedom, and choice). Having too many options to choose from can cause difficulties. A wide range of products badly influences the choice, resulting in less confidence and less satisfaction with the choice. CSR is often overwhelming due to the excessive number of choices. This creates concern about intentions: the barrier to the full implementation of CSR is the fear that CSR efforts will not produce the desired results and be devalued.

Managers have concerns about the motives behind linking several cognitive challenges; fears about intentions may limit managers' and employees' full engagement in CSR. The second type is emotional barriers. In this respect, barriers can be described as fear of evaluating the actions taken. Greater emphasis on CSR will lead to a more comprehensive approach to it and a deeper and more severe assessment; the fear of a negative assessment is demotivating. The fear of scrutiny also exists. In the case of several managers, the adoption of more coordinated measures to address CSR produces a different kind of fear: the fear of a thorough analysis of current CSR programs and activities. Distinguished from the concern, control refers project managers in terms of the need to work on this elements and the need to manage it.

Barriers to the implementation of CSR in organizations can be observed on three distinct levels: individual, organizational, and institutional. Individual-level barriers are psychological and behavioral, and focus on the cognitions of individual decision-makers. Organizational barriers are more structural in nature and focus on issues such as the dominant organization culture, reward systems used, level of organizational inertia, and interactions among individuals. Institutional barriers exist beyond the individual organization level and include macro-rules and regulations, mimetic tendencies of organizations established within norms, and beliefs perpetuated at the institutional level (Garavan et al., 2010). These barriers are described in Table 4.3.

4.3 Corporate Social Responsibility: Trends in Development

The development trends in CSR are related to the general situation off the markets and economies and are associated with the creation of innovative business models in which social or societal aspects are integrated within the organization's strategy. An important element is to make public the information about CSR activities. This may help stakeholders to better understand the philosophy of the organization's operations as well as to indicate the results achieved in terms of sustainable development. Reporting is part of voluntary practices. In this respect, reporting may refer to international standards (GRI) or focus on its guidelines. The scope of reporting may vary depending on the organization and what it wants to demonstrate. A wider scope of reporting is one of the trends expected in the implementation and publication of CSR information.

Table 4.3 Three distinct levels of CSR barriers

Barrier	*Definition*
Individual level	
CSR/CS knowledge awareness	Employee awareness of CSR/CS policies, strategies, and initiatives implemented by the organization plus general environmental awareness
CSR/CS fit, motivation and commitment	Perceived fit of CSR/CS activities with the organizations activities and values. The motivations of the organization for implementing CSR/CS. Motivation and commitment of the employee to implement CSR/CS activities
Perceived organizational and supervisory support	Perceived organizational support focuses on global beliefs concerning whether the organization values the contributions of employees. Perceived supervisory support focuses on the extent to which employees believe that environmental actions are supported by supervisors
Attitudes toward CSR/CS	General attitudes that employees have toward investment in CSR/CS
Personal social actions	Actions supported by and/or sponsored by the organization to support a social good. Specific focus on actions that have involvement of employees
Egocentrism and positive illusions	The potential of people to make self-serving judgments concerning what is fair or considered good. They will as a result make decisions that they perceive to be fair.
Perceptions of organizational justice	The extent to which employees make judgments concerning the fairness of the social concerns and actions of their employing organizations
Organizational level	
Ethical culture and climate	Culture focuses on sense-making devises that are carriers of meaning. Shared meanings concerning the importance of CSR/CS. Climate focuses on the conditions that impact the kinds of values and attitudes that prevail

(continued)

Table 4.3 (continued)

Barrier	*Definition*
Organizational structures	Organization structures focus on roles, responsibilities, patterns of interaction, and authority levels
Teamwork processes	Processes of collaboration, knowledge sharing, and team spirit within an organization
Organizational trust	The willingness of employees to be vulnerable to an organization's actions and policies
Senior management values and leadership	Collective values of senior management and the extent to which decisions are driven by personal values. Leadership focuses on ethical leadership behaviors
Discretionary corporate responsibilities	Those responsibilities that are undertaken by organizations about which society has no clear message to business
Organizational reward systems	Combination of intrinsic and extrinsic rewards that are formally and informally applied to achieve desired behaviors
Organizational inertia	A desire for stability and tendency to resist change driven by habitual routines, resource limitations, and power bases
Corporate hierocracy	The belief that a firm can be something it is not. Distance between what an organization asserts in respect of CSR/CS and what its performance is in these areas
Institutional level	
Regulatory forces	Factors such as laws, regulations, and regulatory bodies that mandate particular courses of action for organizations in respect of CSR/CS
Normative forces	Influences derived from standards, operating practices, best practice suggestions that lead organizations to adopt particular CSR/CS practices
Cognitive influences	Cognitive practices in the form of beliefs shared by organizations that influence the adoption or not of particular practices in organizations

Source Garavan et al. (2010)

When determining both the benefits, barriers, risks, and costs of implementing CSR, the trends that will affect and develop CSR and the overall management of the organization should be considered.

The assumptions and components of the CSR concept allow for the development and setting of new global trends in corporate social responsibility, which can be defined as follows (https://www.forbes.com/sites/timothyjmcclimon/2020/01/02/5-corporate-social-responsibility-trends-to-follow-in-2020/#4474e6987e69):

1. Searching for the truth. This aspect involves transparency and, above all, the sharing of quality, up-to-date, and true information. In the context of global markets, supply chains, and the proliferation of social media, which can both help and, through fake news, lead to problems, organizations and stakeholders must search for the truth. The practical aspect of this approach involves the use of external, neutral resources and auditors who must explain and verify that the information provided is transparent and accurate.
2. Achieving carbon neutrality. Environmental protection and waste elimination activities are being implemented by organizations to various degrees, but simply reducing the environmental impact of a company's activities is not enough; nowadays, stakeholders are demanding that companies eliminate all negative factors. Carbon neutrality should be considered jointly with the idea of zero waste.
3. Standing with employees. Increasingly, employees are using their voices to demand that their corporate leaders get involved in their communities. This is happening in ways never seen before. Cultural and political differences between stakeholders should be bridged and corporate leaders should participate in local and global public debates. The voice of employees can be decisive not only in the implementation of the CSR concept but also in the achievement of other organizational goals.
4. Pursuing purpose over passion. Although passion remains critical for leadership of companies and civic organizations alike, there is a growing recognition that passions are sometimes transient, but purpose can be enduring. The movement toward creating corporate purpose as a way of inspiring employees and engaging customers is a recognition that purpose can be a long-lasting and geographically agnostic guiding force behind a company's operations and its social responsibility.

5. Looking for the next big thing. The concept of corporate social responsibility has existed since the 1970s, and corporate philanthropy has existed for over a hundred years. The idea of sustainable development has been around for decades, and the commitment of companies and employees to give as part of volunteering is equally long-lasting. What is the next big thing? Many investors and insiders favor the idea of environmental, social, and management (ESG) reporting and shared value, but these concepts may seem too academic for the average consumer or employee. Therefore, this aspect can and should become a new challenge for organizations, chains, and components of the strategy.

CSR means being responsible members of communities for a long time. CSR in financial sector includes employee volunteering, mentoring, sponsorship of local activities, and implementation of corporate community investment projects, as well as measuring their outputs and the long-term impact on society and the companies. CSR improves internal processes related to environmental and social management.

The development trends in CSR, related to many issues, have undergone different stages. A simple and illustrative framework for CSR is Carroll's pyramid of CSR (Carroll, 2016). Each of the four components of responsibility involves different stakeholders in terms of the different priorities to which they may have that may impact on stakeholders. Economic responsibility has the greatest impact on shareholders and employees because if a company is not financially viable, both these groups will be significantly affected. Legal obligations are important for owners, but in today's society, the threat of legal proceedings against companies is most often the fault of employees and interested consumers. Ethical obligations affect all stakeholder groups. Stakeholder suits in this area may be a growing concern. Research on ethical issues faced by modern business usually refers to employees, customers, and the environment. The philanthropic responsibilities most affect the community and non-profit organizations as well as employees, as some studies have shown that a company's philanthropic commitment is largely related to the morale and commitment of its employees. The pyramid should be seen as sustainable, since the responsibilities indicated within it are long-term commitments that include future generations of stakeholders. The pyramid can be seen as a static picture of responsibilities, and as a dynamic platform where the framework and its content can be adjusted in the present reality as well as in the future. This is how the classic CSR pyramid is considered.

An important element is the reconfiguration and replacement of the term CSR. Currently, we no longer refer to the issues as corporate social responsibility but corporate sustainability and responsibility (Figs. 4.2 and 4.3).

However, the development of the concept and its evaluation have led to the creation of the CSR 2.0 DNA model (Table 4.4).

The main principles of CSR 2.0 relate to several aspects: C, creativity; S, scalability; R, responsiveness; 2, glocality; and 0, circularity. The success of the new strategy is based on above aspects and principles (Fig. 4.4).

The changes in the understanding of CSR 1.0 and CSR 2.0 are occurring at different levels, as reflected in Table 4.5.

In characterizing CSR 2.0, it refers to and is defined by global commons, innovative partnerships, and stakeholder involvement. Management mechanisms include diverse stakeholder panels, real-time transparent reporting, and new-wave social entrepreneurship, recognizing a shift in power from centralized to decentralized; a change in scale from few and big to many and small, and a change in application from single and exclusive to multiple and shared. CSR 2.0, for implementing organizations, presents a challenge, a vision, and a kind of strategy, with specific and strictly defined objectives. In terms of responsibility, solutions, actions, and management must emerge, so that the often deflected social, economic, and environmental appetites can be found in the new business structure and strategy.

In recent years, the foundations of the CSR 2.0 pyramid have been created (Fig. 4.5). A characteristic feature is that the company must incorporate CSR principles into all its processes. This is related to risk management and strategic CSR being placed at the top of the pyramid as

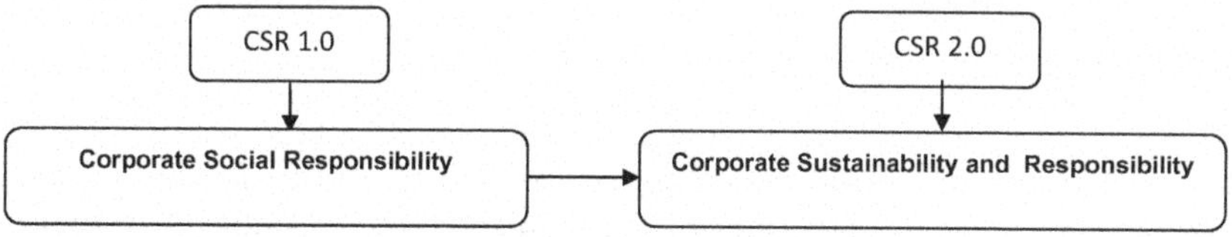

Fig. 4.3 From CSR 1.0 to CSR 2.0 (*Source* Own elaborations)

Table 4.4 CSR 2.0 DNA model

DNA code	*Strategic goals*	*Key indicators*
Value creation	Economic development	Capital investment (financial, manufacturing, social, human and natural capital) Beneficial products (sustainable and responsible goods and services) Inclusive business (wealth distribution, bottom of the pyramid markets)
Good governance	Institutional effectiveness	Leadership (strategic commitment to sustainability and responsibility) Transparency (sustainability and responsibility reporting, government payments) Ethical practices (bribery and corruption prevention, values in business)
Societal contribution	Stakeholder orientation	Philanthropy (charitable donations, provision of public goods and services) Fair labor practices (working conditions, employee rights, health and safety) Supply chain integrity (SME empowerment, labor and environmental standards)
Environment al integrity	Sustainable ecosystems	Ecosystem protection (biodiversity conservation and ecosystem restoration) Renewable resources (tackling climate change, renewable energy and materials) Zero waste production (cradle-to-cradle processes, waste elimination)

Source Visser (2010)

the most important priorities of the organization. In this model, philanthropy is placed at the bottom of the pyramid. This does not mean that it is not important, but its role and character have changed (Visser, 2016).

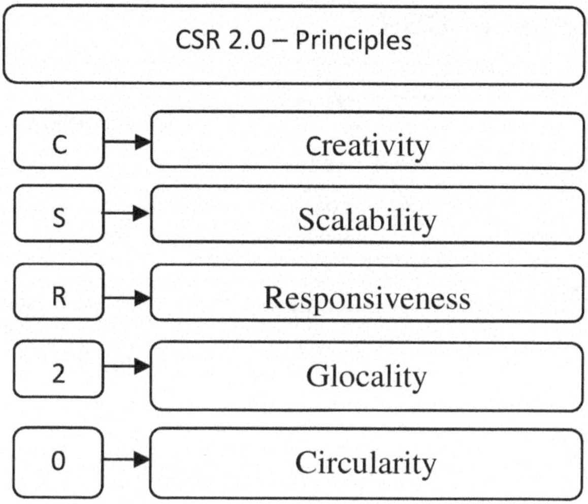

Fig. 4.4 CSR 2.0 principles (*Source* Own elaborations)

Table 4.5 CSR 1.0 to CSR 2.0 meta- and micro-level shifts

Meta-level ontological shifts		*Micro-level methodological shifts*	
CSR 1.0	*CSR 2.0*	*CSR 1.0*	*CSR 2.0*
Philanthropic	Collaborative	CSR premium	Base of the pyramid
Risk-based	Reward-based	Charity projects	Social enterprise
Image-driven	Performance-driven	CSR indexes	CSR ratings
Specialized	Integrated	CSR departments	CSR incentives
Standardized	Diversified	Product liability	Choice editing
Marginal	Scalable	Ethical consumerism	Service agreements
Western	Global	CSR reporting cycles	CSR data streams
		Stakeholder groups	Social networks
		Process standards	Performance standards

Source Visser (2011)

Regarding the future of CSR, businesses will be expected to have the greatest influence on the future dissemination of CSR. Competitors, purchasers, and end consumers (i.e., business stakeholders) will be perceived as the most important brokers of CSR in the future, followed by public authorities and NGOs (i.e., public-oriented stakeholders). Trade

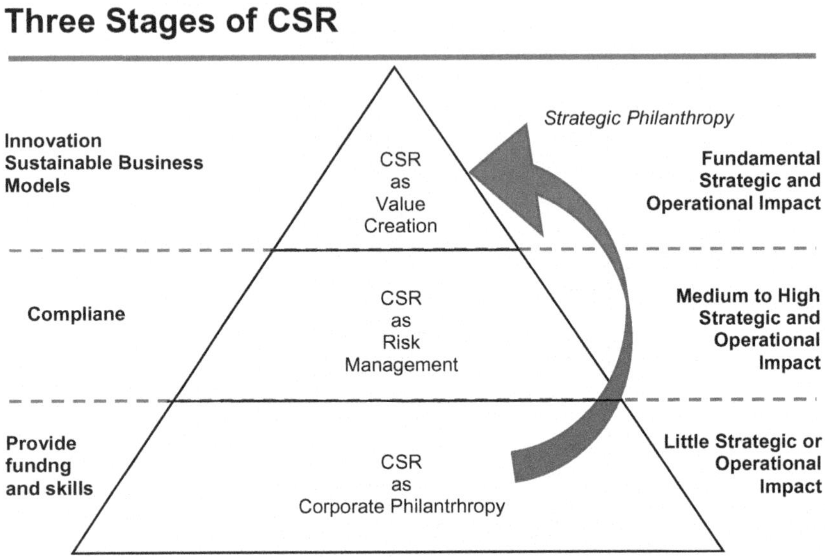

Fig. 4.5 CSR 2.0 pyramid (*Source* Visser [2016])

unions, along with companies from other sectors, are perceived to have the weakest power in this regard (Kudłak et al., 2018). Due to the development trends of societies, sustainable development is nowadays inextricably linked to CSR, so it can also be considered as part of the future trend in CSR. To this must be added responsibility in each area of CSR. CSR can provide the basis for creating new value for organizations based on sustainability.

4.4 Sustainable Value Creation: Identification and the Framework

Sustainable value creation must be based on win-win relationships. Without this approach and without a systemic approach in which benefits must be experienced by all stakeholders, we cannot talk about creating modern business models. Modern and creative business at every stage creates value, and this value must meet the requirement of sustainability.

The creation of any new business model must be based on the concept of value. In terms of sustainable value creation, this involves more than just providing value for the customer and realizing economic value (Yang et al., 2017a); sustainable value refers to the economic, social, and environmental impacts (Hart et al., 2003; Figge & Hahn, 2004) created by the company and its value network (Yang et al., 2017a) and perceived by many stakeholders (Stubbs & Cocklin, 2008). Therefore, the creation of value that is sustainable must be linked to the creation of economic, environmental, and social value (Evans et al., 2017).

In terms of economic value, the creation and realization of value in use for and perceived by the customer must be discussed (convenience and functionality) and in the context of socio-economic well-being and transactional value (monetary, market value, and market access) (Bocken et al., 2014).

Environmental value means the impact of an enterprise on the environment and natural capital (Stubbs & Cocklin, 2008). Social values refer to elements that social units consider valuable, which are linked to social well-being and happiness, and to psychological value elements (den Ouden, 2012). Creating value also means preventing potential negative effects of doing business (Tura et al., 2019). Some business models may focus on creating profits, on social aspects, and on increasing social welfare (Lankoski & Smith, 2018). However, a sustainable business model must have a positive value (Dyllick & Rost, 2017). The models should create economic value for the company without negative effects, while having the potential to create wider net-positive benefits from the environmental and social perspectives (Laukkanen & Tura, 2020).

In sustainability terms, the link between values and the life cycle of products and their management must be sought. Links between life cycle management tools and value creation can be identified through different steps: toolbox, business implementation, business benefits, value drivers, and business value. We proceed from left to right by asking the question "for what purpose?", or "what do we want to achieve?" From right to left, we can ask the question "how?" The value creation framework has to be defined for each company, and, in many cases, the path toward value creation might be longer and more sophisticated than what is presented in Fig. 4.6.

Business models have been developing dynamically, which has led to the creation of business model innovation processes (Wirtz et al., 2016). In recent years, the greatest attention has been paid to connecting and seeking links in the creation of business models and sustainable development. The introduction of sustainable development practices has forced

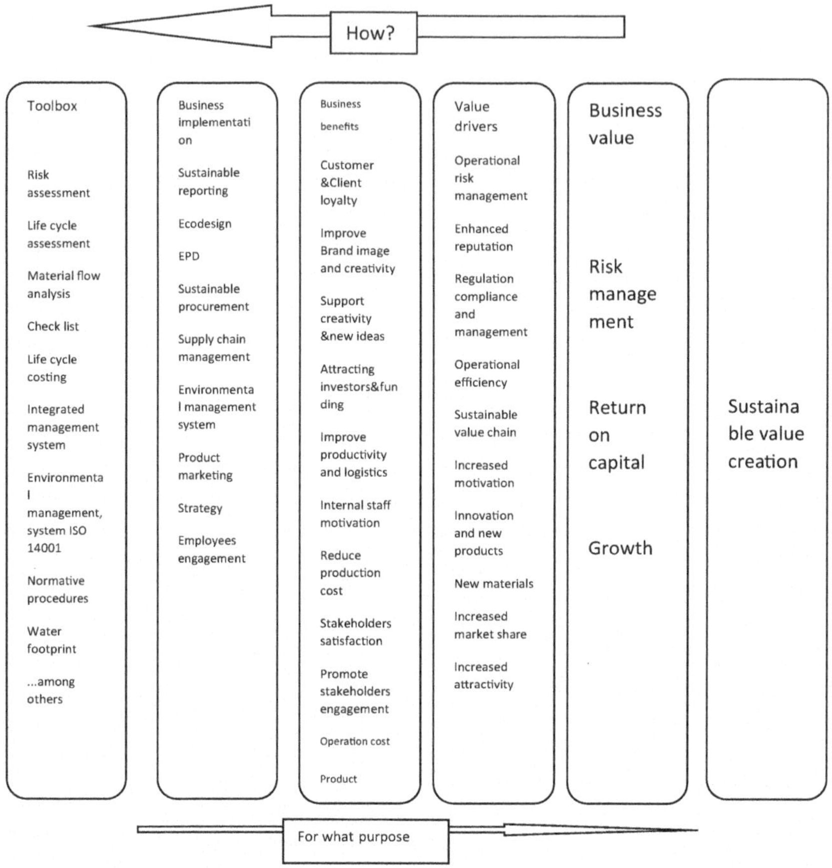

Fig. 4.6 Sustainable value creation—framework (*Source* Sonnemann and Margni [2015])

the redesign not only of business strategy but also the process of creating value. Sustainable business models that allow the implementation of sustainable development principles and innovative solutions that create added value for a given business are therefore increasingly being used (Evans et al., 2017; Yang et al., 2017b).

Employees are needed to create sustainable business models. The triple bottom line (TBL) perspective on value creation integrates the economic, environmental, and social dimensions of sustainable development and facilitates understanding of their interrelationships in many ways (Elkington, 1998). TBL is now seen as an essential part of any company's strategy to integrate sustainability into its operations (Porter & Kramer, 2019).

Value plays a larger role in typical business relationships, which are continuous and based on long-term relationships and mutual interactions (Vandaele & Decouttere, 2013). Establishing such relationships allows for mutual learning, business stability, and improvement of organizational skills (Hurtado Jaramillo et al., 2018). A dialogue between suppliers and customers should also be sought. This plane can create a common space for achieving sustainable development and sustainable business that meets the expectations of all stakeholders (Ballantyne et al., 2011). Changes in the business strategy also require changes in partners in terms of improving the capacity to use resources and the implementation of processes using the links involved (Buxel et al., 2015). Sustainable value should be seen as an element that means integrating environmental, social, and financial objectives into business, together with the needs of many stakeholders. This involves long-term planning; this is the perspective of strategy development (Morioka et al., 2018). Business value can be created by different elements, the most important of which are knowledge, cooperation, inclusiveness, funding, entrepreneurism, policies, diversification, and communications (Fig. 4.7).

In creating a new sustainable business model, an important task is to answer the questions that form a component of the new foundations and creation of a business strategy. A sample portfolio of questions is provided in Table 4.6.

Increasing environmental protection requirements and value-creating social values (Peltola et al., 2016), contributing to the development of the concept of add value creation with sustainable factors. Currently, the concept of value creation includes intangible value elements such as psychological, emotional, and cognitive factors and experiences (benefits and sacrifices) of various stakeholder groups (Cronin et al., 2000). The aim of creating business models that consider environmental values in their strategy is to increase resource efficiency. This can be achieved through the use of renewable resources (Bocken et al., 2014). Environmental value can be created, for instance, through reuse, reproducing products and materials, repurposing materials, eliminating waste, and recycling (Geissdoerfer et al., 2017). It also involves a focus on the

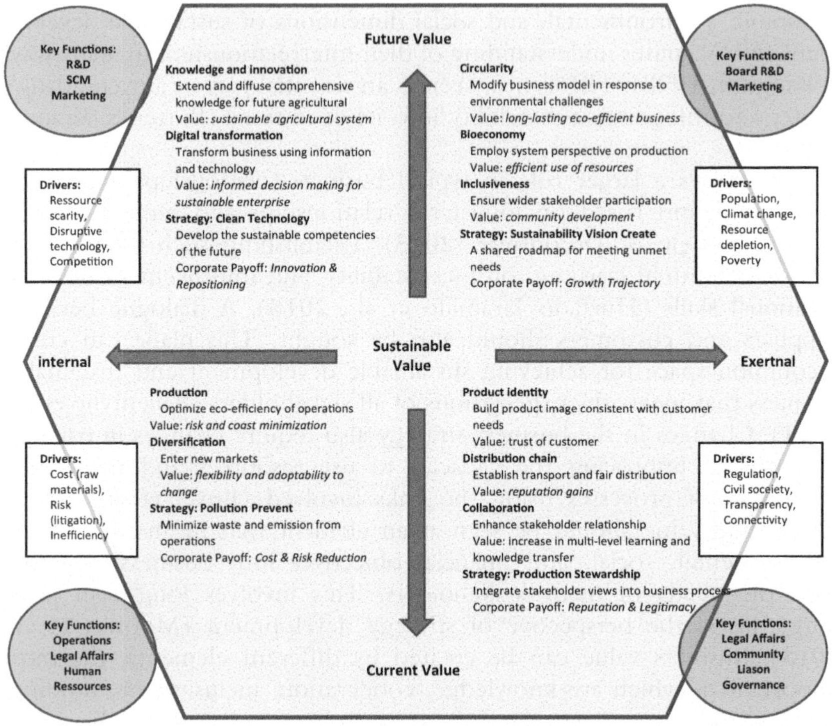

Fig. 4.7 New framework for sustainable value creation (*Source* Sadovska et al. [2020])

responsible use of resources such as water and raw materials and respect for the overall well-being of ecosystems, people, and animals (Rosca et al., 2017).

By definition, products, processes, and services related to sustainable value creation do not have harmful effects on the environment and emissions, such as greenhouse gases, which can harm people or the environment (Bocken et al., 2014). In addition, environmental value can be created by seeking solutions to increase environmental welfare through tackling existing environmental problems or increasing biodiversity.

Solving environmental problems or increasing biodiversity can create environmental value, and through such activities, environmental well-being can be created (Dyck & Silvestre, 2018). When considering value

Table 4.6 The sustainable value portfolio

	Nurturing internal capabilities	*Engaging external constituencies*
Building tomorrow opportunities	*Clean technology* Is the sustainability of products limited by existing competency base? Is there potential to realize major improvements through new disruptive technology? Strategy: deploy the sustainable competencies of the future	*Base of pyramid* Does corporate vision direct toward the solution of social and environmental problems? Does vision focus on serving the unmet needs at the base of the economic pyramid? Strategy: co-create new business for meeting unmet needs
Managing today business	*Pollution prevention* Where are the most significant waste and emission streams from current operations? Can it be lower costs and risk by eliminating waste at the source or by using it as useful input? Strategy: minimize waste and emission from operations	*Product stewardship* What are the implications for product design and development if assume responsibility for a product's entire life cycle? Can it build reputation and legitimacy by engaging a broader range of stakeholders? Strategy: deploy the sustainable competencies of the future

Source Own elaboration on the basis of: Hart (1997)

creation from a social perspective, when creating sustainable business models, the health and safety of employees, customers, and partners, i.e., all stakeholders, must be considered (Bocken et al., 2014; Evans et al., 2017). Ethical and legal aspects should not be forgotten, including non-discrimination, respect for workers' rights, and ethical behavior. All activities, including competition and communication, should be carried out in a way that respects ethical principles and the legal solutions adopted (United Nations. [2019]. About the sustainable development goals. Available at https://www.un.org/sustainabledevelopment/sustainable-development-goals/). The literature also indicates that the aspect of social value should be based on the creation of greater social and psychological well-being (Rosca et al., 2017), as well as overall social and community development (Bocken et al., 2014); increasing social cohesion and mutual interaction is also an important element (Dempsey et al., 2011). In the last area, i.e., creation of economic value (in the context of sustainable business models), the most important aspect is to increase cost-effectiveness. Such actions are possible while increasing productivity, efficiency, and effectiveness in terms of material and energy management (Ambec & Lanoie, 2008), while achieving trade-offs in terms of costs and profitability. The economic efficiency measures taken contribute to conquering new markets or increasing competitiveness (Engert et al., 2016). In general, economic value can be created at the level of society, increasing overall economic prosperity (den Ouden, 2012). Integrating sustainability aspects into an organization's strategy can have long-term benefits, including operational stability and reducing the risk of future uncertainty (Hockerts, 2015). Innovative, sustainable business models can create tangible and measurable business benefits, including influencing the reputation and image of an organization and increasing the attractiveness of an organization, from the perspective of both employees and business partners (Branco & Rodrigues, 2006). Elements of the concept are presented in Table 4.7.

The introduction and adaptation of the economic, environmental, and social aspects of business practices into the organization on which sustainability issues are based must be clearly defined, understandable, and properly communicated to employees, customers, investors, and business partners (Peat, 2003). This understanding of the idea makes it easier to create value in the long term. Activities can be supported by interactions and aligning the value of business entities with the expectations of contractors. Most companies still follow a business value approach that focuses on hardware functionality, performance, market evolution, and financial performance, overlooking other value-generating factors that also affect the overall business balance (Vandaele & Decouttere, 2013).

4.5 Tools for the Creation of Sustainable Value

The scientific literature is rich in studies related to the analysis of frameworks, norms, qualifications, and indices to produce sustainability reports and assess corporate sustainability contributions; however, clear and clear criteria for action to be taken are lacking. This lack of standards for methodology or benchmarks makes it difficult to compare and draw clear conclusions on the use of the analyses in question to study the impact

Table 4.7 Conceptual framework for analyzing sustainable value creation

Environmental	*Social*	*Economic*
Increasing resource efficiency Reuse of products, by-products, and materials. Elimination/reduction of waste. Use of renewables (e.g., energy, raw materials)	Safeguarding health and safety The health and safety of employees/customers/communities are ensured	Increasing cost-efficiency Increases in efficiency and reduced costs compared to alternatives
Responsible use of resources Responsible use of natural resources (e.g., water, raw materials), respecting welfare of ecosystems, people and animals. No creation of rebound effects	Respecting laws, regulations and rights Laws, standards and regulations (e.g., taxes, terms of use) and individuals' rights (e.g., privacy) are respected	Increasing profits and business opportunities Increases in profits and/or creation of new business opportunities and markets
No harmful environmental impacts and emissions No emissions (e.g., greenhouse gases) harming people or the environment. No harm to ecosystems or the environment	Respecting employee, stakeholder and individual rights Employees' and stakeholders' terms are handled fairly (e.g., via paying living wages and non-discrimination). Equal treatment of employees/stakeholders/individuals	Operational stability and risk reduction Increases in long-term stability and risk reductions
Increasing environmental well-being Increases in biodiversity and environmental wellbeing by repairing previous damage and solving environmental problems (e.g., reducing ozone depletion)	Ethical principles and no harmful social impacts Operations, products and services do not harm people or communities. Human rights are respected (e.g., no child labor) Ethical principles are followed (e.g., caring use of resources, honest competition)	Increasing attractiveness Increases in reputation and brand value (e.g., attractiveness as an employee/collaborative partner)
	Increasing social well-being Increases in socio-psychological welfare (e.g., happiness, social cohesion)	Increasing economic well-being Increases in socio-economic welfare (e.g., employment)

Source Laukkanen and Tura (2020)

of implementing sustainable development during the creation of business strategies (Lozano, 2012; Renard, 2015; Hurtado Jaramillo et al., 2018). The proposed tools and the evaluation element are characterized by imperfections, but they are a step toward creating an interpretation describing construction models (Table 4.8). Importantly, companies must consider value creation from multi-stakeholders' views (i.e., customers, suppliers, employees, or even society).

Many scientific studies in recent years have examined the impact of sustainability on business performance, but many gaps remain related to how supplier and customer requirements are aligned with sustainability business interests and how this gap affects the value creation process and its evaluation (Doualle et al., 2015; Zijp et al., 2015). The value mapping tool was developed to assist in sustainable business modeling and aims to assist users (Bocken et al., 2015):

- Understanding the positive and negative aspects of value in a network of stakeholders;
- Identifying conflicting values (i.e., where one stakeholder benefit creates a negative for another stakeholder);
- Identifying opportunities for sustainable business model redesign, especially to improve societal and environmental impact, based on qualitative value judgments.

The value mapping tool takes a multi-stakeholder perspective and considers different forms of value. It considers multiple forms of value: value captured, value missed, value destroyed, and new value opportunities (Bocken et al., 2013). Widely used tools are the sustainable value analysis tool (SVAT) and the Cambridge value mapping tool (CVMT). Value forms in SVAT (Yang et al., 2014) include value captured (VC), value absence (VA), value surplus (VS), value destroyed (VD), value missed (VM), and value opportunity (VO). Figure 4.8 explains the definitions of the value forms and the process of value analysis in SVAT. Among the value forms, the value destroyed and value missed are adapted from the value mapping tool; value absence and value surplus are newly proposed to enhance the development of new business strategies. SVAT groups the value forms into three parts (value captured, value uncaptured, and value opportunity) by merging the value absence, value surplus, value

Table 4.8 Tools and framework in sustainability value concepts for business

Author(s)	*Description*
Hart, S. L., & Milstein, M. B. (2003). Creating sustainable value. *Academy of Management Perspectives*, *17*(2), 56–67	First development of a framework for the concept of sustainable development values for business. An attempt to answer the question of how companies implement sustainable development principles in their business strategy. They suggest its use by managers to determine business activity in each of the four quadrants described in their model in order to identify imbalance in its portfolio of activities. The framework was built taking as base global drivers and challenges of sustainability to identify strategies and practices that contribute to achieve simultaneously both SDG and shareholder value. For using this tool, the manager's first step toward the creation of sustainable value within the business is thinking through a wide and full range of challenges and opportunities. Model is based on the activities organization across discrete projects and business experiments
Juwana, I., Perera, B. J. C., & Muttil, N. (2010). A water sustainability index for West Java. Part 1: Developing the conceptual framework. *Water Science and Technology*, *62*(7), 1629–1640. Juwana, I., Perera, B. J. C., & Muttil, N. (2010). A water sustainability index for West Java–Part 2: refining the conceptual framework using Delphi technique. *Water Science and Technology*, *62*(7), 1641–1652	Authors developed a conceptual framework of water sustainability index for West Java(WJWSI). The method is composed by three main parts: components, indicators/sub-indicators, and threshold values. For built framework there were invited lecturers, government officials, consultants and community representatives in their role of water-related stakeholders. As result, the water sustainability index provides information on the current conditions of water resources, as well as the priority of water issues. Although WJWSI is a powerful tool for water utilities, its limitations to highlight value creation process in business are evident

(continued)

Table 4.8 (continued)

Author(s)	*Description*
Yang, M., Vladimirova, D., Rana, P., & Evans, S. (2014). Sustainable value analysis tool for value creation. *Asian Journal of Management Science and Applications*, *1*(4), 312–332	Propose a sustainable value analysis tool (SVAT) in order to help manufacturers identify opportunities to create sustainable value by means to analyze 4 specific aspects: product life cycle, value proposition, uncaptured value, and value opportunity. SVAT combines the life cycle thinking and the analysis of different value forms
Dr. Samuel Short, Dr. Nancy Bocken, Dr. Dai Morgan, Dr. Lloyd Fernando, Dr. Curie Park, Dr. Fenna Blomsma and Dr. Maria Holgado in the scope of "*EU FP7 Sustain Value project*"	Authors create The Cambridge Value Mapping Tool (CVMT). This last tool was built recently, and it has been developed to identify failed value exchanges among multi-stakeholders in the firm network uncovering new value opportunities. CVMT having accounted for the analysis of different value forms
Evans, S., Fernando, L., & Yang, M. (2017). Sustainable value creation—from concept toward implementation. In *Sustainable Manufacturing* (pp. 203–220). Springer, Cham	Authors describe and analyze in their study the strengths and weakness of the CVMT, and the SVAT

Source Hurtado Jaramillo et al. (2018)

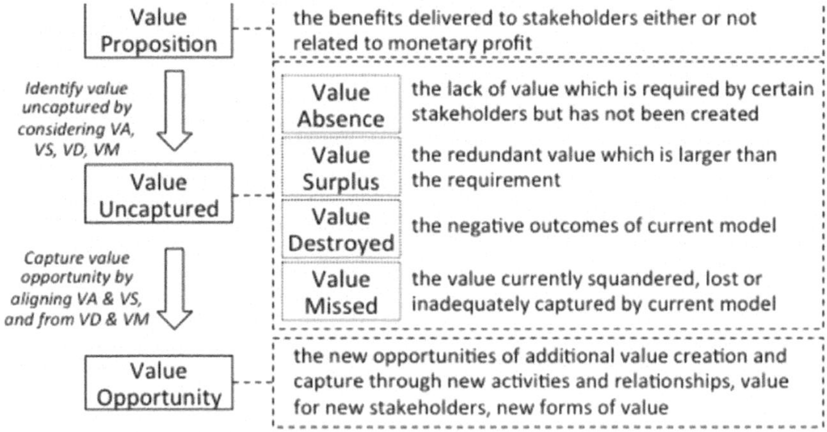

Fig. 4.8 Multiple value forms in SVAT (*Source* Yang et al. [2014])

destroyed, and value missed into value uncaptured (VU). VC is the benefits delivered to the company and its stakeholders, VU is the value that is not captured by the company but has the potential to generate value opportunities, and VO is the new opportunities for value creation. The identification of VC and VU can assist companies in understanding the positive and negative aspects of their current business models. The analysis of VU can assist companies in identifying VO. However, the identification of VU in service-oriented companies is difficult since they are usually hidden and invisible in service process. The rationale of SVAT is to use separate forms of value (i.e., VA, VS, VD, and VM) to inspire the identification of VU, and to further identify VO by analyzing the identified VU (e.g., aligning VA and VS, and reducing VD and VM).

The SVAT can be combined with different product life cycles. The tool is presented in Table 4.9. The tool combines life cycle thinking and value forms analysis. The three phases of a product life cycle (BOL, MOL, and EOL) could be further divided into more specific stages. For example, MOL can be further divided into distribution, use, maintenance, and service. The value forms consist of value captured, value uncaptured, and value opportunity. Value uncaptured could be considered from the perspectives of value destroyed, value missed, value surplus, and value absence.

Table 4.9 SVAT tool behalf on life cycles

Sustainable value analysis tool (SVAT)		*Beginning of life (BOL)* *e.g. Planning, design, procurement, production*	*Middle of life (MOL)* *e.g. Distribution (retail), use, maintenance, service*	*End of life (EOL)* *e.g. Recycle, remanufacture, reuse, disposal*
Value captured (VC)		BOL-VC	MOL-VC	EOL-VC
Value uncaptured (VU)	Value destroyed Value missed Value surplus Value absence	BOL-VU	MOL-VU	EOL-VU
Value opportunity (VO)		BOL-VO	MOL-VO	EOL-VO

Source Yang et al. (2014)

The The Cambridge Value Mapping Tool (CVMT) has been developed and refined over the last years. It uses a structured and visual approach to identify value uncaptured in the form of failed value exchanges: value missed, destroyed, surplus, and absence. The exchange of value is analyzed through the lens of each stakeholder in the business network, with the natural environment and society each being given its own voice and stake in the business. The tool provides practitioners with a new method to gain a deeper understanding of value and create new economic, social, and environmental benefits for their business. The CVMT provides step-by-step process to answer the following questions:

1. What is the unit of analysis of, e.g., the product, service, company, or industry?
2. Who are the stakeholders of the unit of analysis?
3. What is the purpose of the unit of analysis?
4. What is the current value captured?
5. What is the value missed and/or destroyed?
6. What is the value surplus and/or absence?
7. What are the new value opportunities?

The tool is visualized in Fig. 4.9.

Development of the concept of CSR is linked to the creation and implementation of new innovative business models. To outline the process of value creation, the business model has been considered for years as an essential tool in business activity. Business models have developed on both theoretical and practical levels. The most commonly used tool is the Business Model Canvas (BMC), which was first proposed by Osterwalder et al. (2010). BMC is simple and intuitive. Specifically, the BMC describes through nine blocks (key partners, key activities, key resources, value proposition, customer relationships, customer segments, channels, cost structure, and revenue streams), logically demonstrating how a company expects to generate profit. The value proposition is the viewpoint of the BMC, which explains the benefit that the company creates for the client and why they should choose that company among many others (García-Muiña, F. E., Medina-Salgado, M. S., Ferrari, A. M., and Cucchi, M. [2020]. Sustainability transition in industry 4.0 and smart manufacturing with the triple-layered business model canvas. *Sustainability*, 12[6], 2364; Gassmann, O., Frankenberger, K., and Csik, M. [2014].

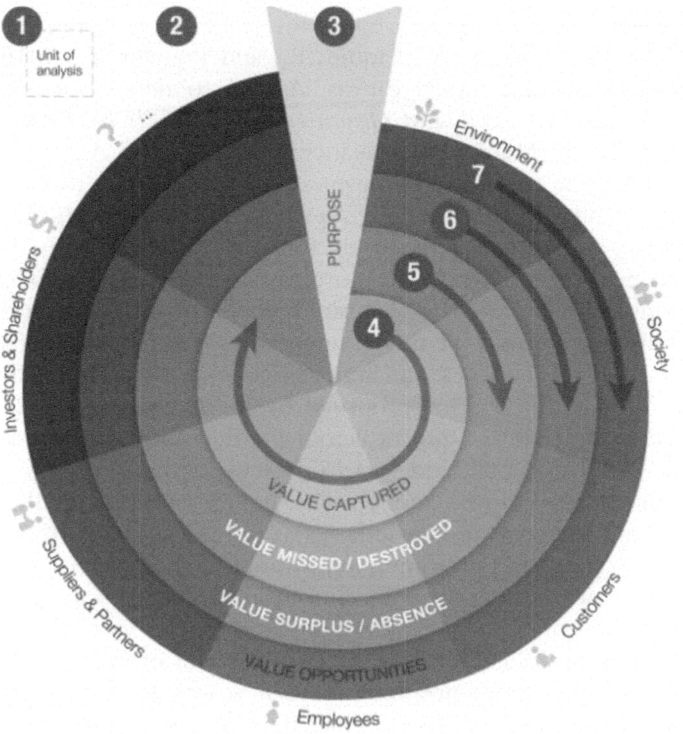

Fig. 4.9 Visualization of the Cambridge value mapping tool (*Source* https://www.ifm.eng.cam.ac.uk/research/industrial-sustainability/sustainable-business-models/tools/cambridge-value-mapping-tool/)

The Business Model Navigator: 55 Models That Will Revolutionize Your Business). Pearson presented a model representation based on four dimensions: who, what, how, and why. The model is represented as a triangle where the client is in the center (who). In the corners, there are the value proposition (what), value chain (how), and profit mechanism (why). These three aspects are placed at the corners of the triangle because modifying one of them therefore also requires modifying the other two (Gassmann, O., Frankenberger, K., and Csik, M. [2010]. *The Business Model Navigator*. London, UK: Pearson).

The development of the concept is the triple-layered business model canvas (TLBMC) (Joyce, A., Paquin, R., and Pigneur, Y. [2015]. The triple layered business model canvas: A tool to design more sustainable business models. *ARTEM Organizational Creativity International Conference*, 26–27 March 2015, Nancy, France; Joyce, A., and Paquin, R. L. [2016]. The triple layered business model canvas: A tool to design more sustainable business models. *Journal of Cleaner Production*, 135, 1474–1486). It is one of the tools for analyzing innovative business models to help create and design more sustainable business models. The model consists of three separate business models, one for each pillar of sustainability. As a result, the model integrates the three pillars of sustainable development and does not limit the analysis of value propositions to the economic aspect. First proposed by Joyce and Paquin (2016), it should be considered as a practical tool for the coherent integration of economic, environmental, and social problems within a holistic view of the organization's business model. The model is based on the classic CANVAS model, providing practical support to the creative development, visualization, and informing innovations in the sustainable business model. As a tool, it combines the innovation (Spieth et al., 2014) and implementation of sustainable development aspects into the business model (Boons & Lüdeke-Freund, 2013). The aim of implementing this model is to support organizations in their search for opportunities to implement changes related to the challenges of sustainable development (Shrivastava, P., and Statler, M. [2012]. *Learning from the Global Financial Crisis: Creatively, Reliably, and Sustainably*. Stanford University Press). Visualizations of the model's planes are presented in Fig. 4.10.

TLBMC is a tool to support the creative search for sustainable business models and, more broadly, sustainable innovation. The TLBMC provides horizontal coherence within each canvas layer for exploring economic, environmental, and social value individually, and vertical coherence integrating value creation across the three canvas layers, which supports developing a deeper understanding of an organization's value creation (Lozano, 2008). The solutions contained in the model can help overcome the barriers associated with the implementation of sustainable development principles into business reality. Environmental aspects and sustainability are elements and creators of innovation, while creating new value for all stakeholders. This new tool for analyzing business models helps to identify the interrelationships and relationships between an organization's current operations and social, environmental, and economic

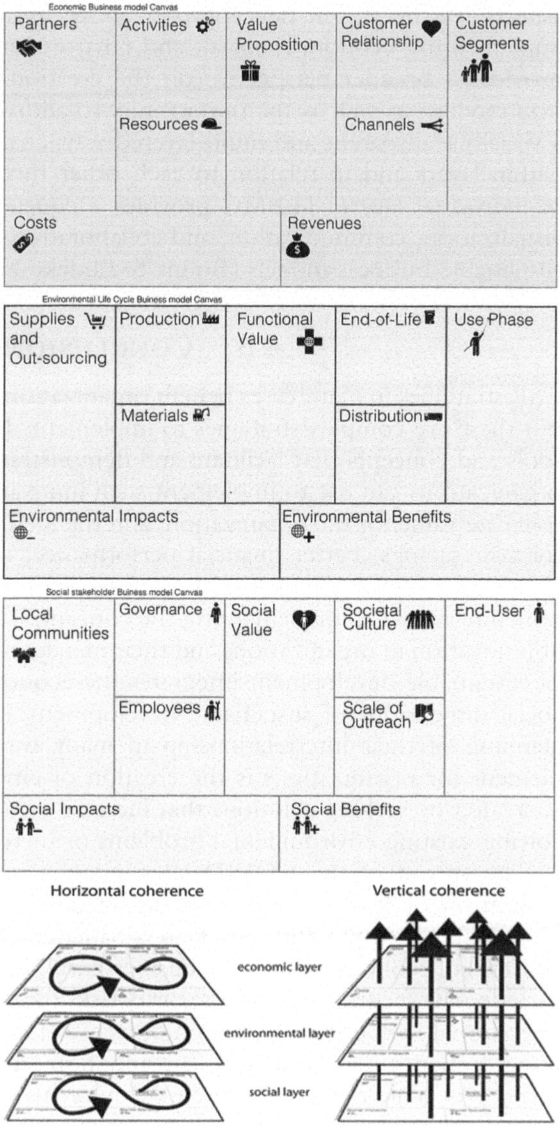

Fig. 4.10 TLBMC visualization (*Source* Joyce and Paquin [2016])

aspects in general. The new approach allows for a more holistic approach, implementing economic, social, and environmental value. The TLBMC provides a broader perspective on the creation and evaluation of business models, as well as the opportunity to address sustainability issues in a visually transparent and multi-layered way, enabling horizontal research within layers and in relation to each other through vertical integration of individual layers. TLBMC provides a concise framework to support visualization, communication, and collaboration around innovating more sustainable business models (Boons & Lüdeke-Freund, 2013).

4.6 Conclusion

CSR strategies in many cases benefit organizations and their supply chains, but these are complex strategies to implement. However, there are many tools and concepts that facilitate and demonstrate the best solutions that organizations can use to implement such innovative solutions. CSR helps to create value for the organization, and the identified driving forces allow for cost savings, better financial performance, or a better image of the organization on the market. However, the barriers and problems must be remembered when implementing the concept. The perspective of looking at innovation in organizations and their management from the perspective of sustainable development integrates the economic, environmental, and social dimensions of sustainable development and facilitates the understanding of their interrelationship in many ways. Today, an important element for organizations is the creation of environmental value, which is created by seeking solutions that increase environmental well-being by solving existing environmental problems or increasing biodiversity.

The spread of the COVID-19 pandemic in 2020 also affected the realization of CSR objectives in organizations. In the context of CSR areas, the pandemic affected (García-Sánchez & García-Sánchez, 2020) the level of economic risk at the business level. Key elements of this impact include new security and cybersecurity risks for employees and customers, operational risks due to limitation of face-to-face economic activity, operating costs not correlated with income, liquidity problems, and survival. Socio-economic risk can be discussed in terms of unemployment, loss of income, the emergence of situations of vulnerability, massive appearance of psychological problems due to pain from the loss of loved ones or from problems associated with the new personal and work situations, feelings of loneliness, and the need for training and leisure activities at home due to limited mobility.

A pandemic contributes to redefining business strategies. Pressures and global trends affecting the implementation of CSR concepts in organizations must now consider all the effects of COVID-19. It is currently impossible to predict the economic impact of the pandemic in the future, but restrictions on business activities have negatively affected companies. It is therefore necessary to consider in the future how CSR practices, due to the need to incur financial expenses, which do not always quickly translate into visible environmental or social effects, should be implemented, so that even in crisis situations, their implementation does not cease. In its initial phase, the COVID-19 pandemic has already demonstrated the existing problems (supply of, e.g., medical equipment); therefore, the aspects of social support and responsible involvement of organizations and their support (free provision of equipment for use, assistance in providing equipment by companies) may prevent such situations while realizing the mission of CSR.

Literature

Agudo-Valiente, J. M., Garcés-Ayerbe, C., & Salvador-Figueras, M. (2017). Corporate social responsibility drivers and barriers according to managers' perception: Evidence from Spanish firms. *Sustainability, 9*(10), 1821.

Ambec, S., & Lanoie, P. (2008). Does it pay to be green? A systematic overview. *Academy of Management Perspectives, 22*, 45–62.

Ballantyne, D., Frow, P., Varey, R. J., & Payne, A. (2011). Value propositions as communication practice: Taking a wider view. *Industrial Marketing Management, 40*(2), 202–210.

Bian, J., Liao, Y., Wang, Y. Y., & Tao, F. (2020). Analysis of firm CSR strategies. *European Journal of Operational Research, 290*(3), 914–926.

Bocken, N. M. P., Rana, P., & Short, S. W. (2015). Value mapping for sustainable business thinking. *Journal of Industrial and Production Engineering, 32*(1), 67–81.

Bocken, N., Short, S., Rana, P., & Evans, S. (2013). A value mapping tool for sustainable business modelling. *Corporate Governance, 13*, 482–497.

Bocken, N., Short, S., Rana, P., & Evans, S. (2014). A literature and practice review to develop sustainable business model archetypes. *Journal of Cleaner Production, 65*, 42–56.

Boons, F., & Lüdeke-Freund, F. (2013). Business models for sustainable innovation: State-of-the-art and steps towards a research agenda. *Journal of Cleaner Production, 45,* 9–19.

Branco, M. C., & Rodrigues, L. L. (2006). Corporate social responsibility and resource based perspectives. *Journal of Business Ethics, 69,* 111–132.

Bux, H., Zhang, Z., & Ahmad, N. (2020). Promoting sustainability through corporate social responsibility implementation in the manufacturing industry: An empirical analysis of barriers using the ISM-MICMAC approach. *Corporate Social Responsibility and Environmental Management, 27*(4), 1729–1748.

Buxel, H., Esenduran, G., & Griffin, S. (2015). Strategic sustainability: Creating business value with life cycle analysis. *Business Horizons, 58*(1), 109–122.

Carroll, A. B. (1991). The pyramid of corporate social responsibility: Toward the moral management of organizational stakeholders. *Business Horizons, 34*(4), 39–48.

Carroll, A. B. (2016). Carroll's pyramid of CSR: Taking another look. *International Journal of Corporate Social Responsibility, 1*(1), 1–8.

Chan, H. L., Wei, X., Guo, S., & Leung, W. H. (2020). Corporate social responsibility (CSR) in fashion supply chains: A multi-methodological study. *Transportation Research Part E: Logistics and Transportation Review, 142,* 102063.

Chandler, D. (2019). *Strategic corporate social responsibility: Sustainable value creation.* Sage, Incorporated.

Cronin, J. J., Brady, M. K., & Hult, G. T. M. (2000). Assessing the effects of quality, value and customer satisfaction on consumer behavioral intentions in service environment. *Journal of Retailing, 76*(2), 193–218.

Dahlsrud, A. (2008). How corporate social responsibility is defined: An analysis of 37 definitions. *Corporate Social Responsibility and Environmental Management, 15*(1), 1–13.

Dempsey, N., Bramley, G., Power, S., & Brown, C. (2011). The social dimension of sustainable development: Defining urban social sustainability. *Sustainable Development, 19*(5), 289–300.

den Ouden, E. (2012). *Innovation design: Creating value for people, organizations and society.* Springer.

Dixit, A. (2009). Governance institutions and economic activity. *American Economic Review, 99*(1), 5–24.

Doualle, B., Medini, K., Boucher, X., & Laforest, V. (2015). Investigating sustainability assessment methods of product-service systems. *Procedia Cirp, 30,* 161–166.

Dyck, B., & Silvestre, B. S. (2018). Enhancing socio-ecological value creation through sustainable innovation 2.0: Moving away from maximizing financial value capture. *Journal of Cleaner Production, 171,* 1593–1604.

Dyllick, T., & Rost, Z. (2017). Towards true product sustainability. *Journal of Cleaner Production, 162,* 346–360.

Elkington, J. (1998). *Cannibals with Forks: The triple bottom line of 21st century business.* New Society Publishers. ISBN 978-0-86571-392-5.

Engert, S., Rauter, R., & Baumgartner, R. J. (2016). Exploring the integration of corporate sustainability into strategic management: A literature review. *Journal of Cleaner Production, 112,* 2833–2850.

Evans, S., Vladimirova, D., Holgado, M., Van Fossen, K., Yang, M., Silva, E. A., et al. (2017). Business model innovation for sustainability: Towards a unified perspective for creation of sustainable business models. *Business Strategy and the Environment, 26*(5), 597–608.

Figge, F., & Hahn, T. (2004). Sustainable value added measuring corporate contributions to sustainability beyond eco-efficiency. *Ecological Economics, 48*(2), 173–187.

Garavan, T. N., Heraty, N., Rock, A., & Dalton, E. (2010). Conceptualizing the behavioral barriers to CSR and CS in organizations: A typology of HRD interventions. *Advances in Developing Human Resources, 12*(5), 587–613.

García-Muiña, F. E., Medina-Salgado, M. S., Ferrari, A. M., & Cucchi, M. (2020). Sustainability transition in industry 4.0 and smart manufacturing with the triple-layered business model canvas. *Sustainability, 12*(6), 2364.

García-Sánchez, I. M., & García-Sánchez, A. (2020). Corporate social responsibility during COVID-19 pandemic. *Journal of Open Innovation: Technology, Market, and Complexity, 6*(4), 126.

Gassmann, O., Frankenberger, K., & Csik, M. (2010). *The business model navigator.* Pearson.

Gassmann, O., Frankenberger, K., & Csik, M. (2014). *The business model navigator: 55 models that will revolutionise your business.* Pearson UK 2010.

Geissdoerfer, M., Savaget, P., Bocken, N. M. P., & Hultink, E. J. (2017). The circular economy a new sustainability paradigm? *Journal of Cleaner Production, 143,* 757–768.

Goyal, P., & Kumar, D. (2017). Modeling the CSR barriers in manufacturing industries. *Benchmarking: An International Journal, 24*(7), 1871–1890.

Hart, S. (1997, January–February). Beyond greening: Strategies for a sustainable world. *Harvard Business Review,* 66–76.

Hart, S. L., Milstein, M. B., & Caggiano, J. (2003). Creating sustainable value. *The Academy of Management Executive, 17*(2), 56–67.

Hockerts, K. (2015). A cognitive perspective on the business case for corporate sustainability. *Business Strategy and the Environment, 24,* 102–122.

https://www.forbes.com/sites/timothyjmcclimon/2020/01/02/5-corporate-social-responsibility-trends-to-follow-in-2020/#4474e6987e69.

https://www.ifm.eng.cam.ac.uk/research/industrial-sustainability/sustainable-business-models/tools/cambridge-value-mapping-tool/.

Hurtado Jaramillo, C. H., Arimany Serrat, N., Meijide Vidal, D., & Ferràs Hernández, X. (2018). Sustainability values for business: A perspective of value alignment in a supplier–client relationship for case aqualogy. *Intangible Capital, 14*(1), 3–24.

Joyce, A., & Paquin, R. L. (2016). The triple layered business model canvas: A tool to design more sustainable business models. *Journal of Cleaner Production, 135,* 1474–1486.

Joyce, A., Paquin, R., & Pigneur, Y. (2015, March 26–27). The triple layered business model canvas: A tool to design more sustainable business models. In *ARTEM organizational creativity international conference*. Nancy, France.

Księżak, P. (2016). The benefits from CSR for a company and society. *Journal of Corporate Responsibility and Leadership, 3*(4), 53–65.

Kudłak, R., Szőcs, I., Krumay, B., & Martinuzzi, A. (2018). The future of CSR-selected findings from a Europe-wide Delphi study. *Journal of Cleaner Production, 183,* 282–291.

Lankoski, L., & Smith, N. C. (2018). Alternative objective functions for firms. *Organization & Environment, 31*(3), 242–262.

Laudal, T. (2011). Drivers and barriers of CSR and the size and internationalization of firms. *Social Responsibility Journal, 7*(2), 234–256.

Laukkanen, M., & Tura, N. (2020). The potential of sharing economy business models for sustainable value creation. *Journal of Cleaner Production, 253,* 120004.

Lozano, R. (2008). Envisioning sustainability three-dimensionally. *Journal of Cleaner Production, 16*(17), 1838–1846.

Lozano, R. (2012). Towards better embedding sustainability into companies' systems: An analysis of voluntary corporate initiatives. *Journal of Cleaner Production, 25,* 14–26.

McWilliams, A., & Siegel, D. (2000). Corporate social responsibility and financial performance: Correlation or misspecification? *Strategic Management Journal, 21*(5), 603–609.

Morioka, S. N., Bolis, I., & Carvalho, M. M. D. (2018). From an ideal dream towards reality analysis: Proposing sustainable value exchange matrix (SVEM) from systematic literature review on sustainable business models and face validation. *Journal of Cleaner Production, 178,* 76–88.

Osterwalder, A., Pigneur, Y., & Clark, T. (2010). *Business model generation*. Wiley.

Peat, R. O. S. S. (2003). Values drive value. *University of Auckland Business Review, 5*(2), 1–11.

Peltola, T., Aarikka-Stenroos, L., Viana, E., & Mcakinen, S. (2016). Value capture in business ecosystems for municipal solid waste management: Comparison between two local environments. *Journal of Cleaner Production, 137,* 1270–1279.

Porter, M. E., & Kramer, M. R. (2019). Creating shared value. In G. G. Lenssen & N. C. Smith (Eds.), *Managing sustainable business: An executive education case and textbook* (pp. 323–346). Springer. ISBN 978-94-024-1144-7.

Renard, Y. J. S. (2015). A review of corporate sustainability reporting tools (SRTs). *Journal of Environmental Management, 164*, 180–195.

Reynolds, S. J. (2019). Overcoming the psychological barriers of CSR. *Organizational Dynamics*, No. 1/100745. ISSN 0090-2616. https://doi.org/10.1016/j.orgdyn.2019.100745.

Rosca, E., Arnold, M., & Bendul, J. C. (2017). Business models for sustainable innovation an empirical analysis of frugal products and services. *Journal of Cleaner Production, 162*, 133–145.

Sadovska, V., Ekelund Axelson, L., & Mark-Herbert, C. (2020). Reviewing value creation in agriculture—A conceptual analysis and a new framework. *Sustainability, 12*(12), 5021.

Sarkar, S., & Searcy, C. (2016). Zeitgeist or chameleon? A quantitative analysis of CSR definitions. *Journal of Cleaner Production, 135*, 1423–1435.

Shrivastava, P., & Statler, M. (Eds.). (2012). *Learning from the global financial crisis: Creatively, reliably, and sustainably*. Stanford University Press.

Sonnemann, G., & Margni, M. (2015). *Life cycle management* (p. 353). Springer Nature.

Spieth, P., Schneckenberg, D., & Ricart, J. E. (2014). Business model innovation–state of the art and future challenges for the field. *R&D Management, 44*(3), 237–247.

Stubbs, W., & Cocklin, C. (2008). Conceptualizing a "sustainability business model". *Organization & Environment, 21*(2), 103–127.

Tura, N., Keraanen, J., & Patala, S. (2019). The darker side of sustainability: Tensions from sustainable business practices in business networks. *Industrial Marketing Management, 77*, 221–223.

United Nations. (2019). *About the sustainable development goals*. Available at https://www.un.org/sustainabledevelopment/sustainable-development-goals/.

Van Wassenhove, L. N. (2019). Sustainable innovation: Pushing the boundaries of traditional operations management. *Production and Operations Management, 28*(12), 2930–2945.

Vandaele, N. J., & Decouttere, C. J. (2013). Sustainable R&D portfolio assessment. *Decision Support Systems, 54*(4), 1521–1532.

Villena, V. H. (2019). The missing link? The strategic role of procurement in building sustainable supply networks. *Production and Operations Management, 28*(5), 1149–1172.

Visser, W. (2010). The age of responsibility: CSR 2.0 and the new DNA of business. *Journal of Business Systems, Governance and Ethics, 5*(3), 7.

Visser, W. (2011). *The age of responsibility: CSR 2.0 and the new DNA of business*. John Wiley & Sons.

Visser, W. (2016). The future of CSR: Towards transformative CSR, or CSR 2.0. In *Research handbook on corporate social responsibility in context*. Edward Elgar Publishing.

Wirtz, B. W., Pistoia, A., Ullrich, S., & Gottel, V. (2016). Business models: Origin, development and future research perspectives. *Long Range Plan, 49*(1), 36–54.

Yadlapalli, A., Rahman, S., & Gunasekaran, A. (2020). Corporate social responsibility definitions in supply chain research: An ontological analysis. *Journal of Cleaner Production, 277*,.

Yang, M., Evans, S., Vladimirova, D., & Rana, P. (2017a). Value uncaptured perspective for sustainable business model innovation. *Journal of Cleaner Production, 140,* 1794–1804.

Yang, M., Vladimirova, D., & Evans, S. (2017b). Creating and capturing value through sustainability: The sustainable value analysis tool a new tool helps companies discover opportunities to create and capture value through sustainability. *Research-Technology Management, 60*(3), 30–39.

Yang, M., Vladimirova, D., Rana, P., & Evans, S. (2014). Sustainable value analysis tool for value creation. *Asian Journal of Management Science and Applications, 1*(4), 312–332.

Zhang, D., Morse, S., & Kambhampati, U. (2017). *Sustainable development and corporate social responsibility*. Routledge.

Zhang, Q., Oo, B. L., & Lim, B. T. H. (2019). Drivers, motivations, and barriers to the implementation of corporate social responsibility practices by construction enterprises: A review. *Journal of Cleaner Production, 210,* 563–584.

Zijp, M. C., Heijungs, R., Van der Voet, E., Van de Meent, D., Huijbregts, M. A., Hollander, A., et al. (2015). An identification key for selecting methods for sustainability assessments. *Sustainability, 7*(3), 2490–2512.

CHAPTER 5

Role of Banks in Ensuring Sustainability in Companies Sector

5.1 Introduction

The banks usually have no (or negligible) impact on natural environment, but they understood that they had to make decisions to be more proactive in this field. However, in order to achieve a more sustainable form of bank developing, there is a need for a more holistic perspective that allows to consider all of resources upon which their business relies. Banks learnt that adherence to environmental issues allows them to uncover latent, potentially harmful risks, save money and simultaneously improve their social responsibility reputation. This is a situation when a private sustainability was associated also with improving social sustainability. Banks will have to be responding to the challenges of sustainability, so as to be able to build and finance environmentally friendly activities in the business sector through it, and to change their attitudes through cooperation with the customers to a more environmental activities dedicated to joint achievement of goals sustainable development.

Historically banks have been viewed solely as financial institutions, which should concern themselves with all things financial and there was no environmental factor in their activities. Currently, banks are moving away from the traditional perception of their role toward sustainability. (Liston-Heyes & Ceton, 2007; Pellicano et al. 2014; Leal Filho & Brandli, 2016; Zimmermann, 2019)This public view has allowed banks significant leeway with concern to ethical standards and CSR application.

© The Author(s), under exclusive license to Springer Nature Switzerland AG 2021

M. Ziolo et al., *Sustainability in Bank and Corporate Business Models*, Palgrave Studies in Impact Finance,
https://doi.org/10.1007/978-3-030-72098-8_5

On the one hand, a special place in greening banks and shaping the principles of social responsibility is taken by central banks that participate in pro-environmental international organizations that develop sustainability standards (Carney, 2015; Kiernan, 2014; Leaton et al., 2013; Macklem, 2011; Park & Kim, 2020; Volz, 2017).

On the other hand, banks are looking for new directions of activity and this gives them sustainability (Bansal & Roth, 2000; Cassells & Lewis, 2011; Zaman, 2012; Ferenhof et al., 2014; Pellicano et al., 2014; Johnson & Schaltegger, 2016; Hughes & Troy, 2017). They gain new customers who need support in their environmental activities. There is also a whole set of factors changing banks' business strategies, which is reflected in numerous works in the literature on the subject (Cetindamar & Husay, 2007; Jha & Bhome, 2013; Mishra, 2013; Biswas, 2011; Tara et al., 2015; Wu & Cheng, 2018). By changing their policy, banks create new business models.

The question arises whether banks are able to change the needs in terms of the offer of services and products of their clients? The literature on the subject seeks answers to the question not only about the role of the central bank in changing the business models of commercial banks (ONZ, 2015; Dikau & Ryan-Collins, 2017; Volz, 2017), but also answers to the question about the stability of commercial banks and their commitment to social responsibility and adequacy of the offer toward sustainability (Becchetti et al., 2011; San José et al., 2011; Cornée et al. 2016; Weber, 2016; Park & Kim, 2020).

The chapter is based on both primary and secondary sources. In order to know the current prospects of sustainable bank products and sustainable instruments of central bank, a primary survey was carried out using a case study on policy central banks (which is available on websites) and the offers of green products and the motives for their introduction, which are in the bank's declarations on their websites. Selected bank CSR reports were also analyzed. The facts have been presented after the discussion with colleagues. The secondary data were collected on perusal of the selected available research work, textbook, articles, journals, newspaper analysis, related websites, and analysis of reported case decisions.

Undoubtedly, for the future, research will be important in the search for an answer to the question: whether the measures taken toward sustainability and toward changing business models have brought results. Was the policy of the central bank effective and was it able to influence the desired changes in the business models of a commercial bank? Answers to these questions should be searched for in the future, as the longer perspective will help to improve the answers.

5.2 Corporate Motives for Sustainability

Along with the growing social, environmental, and economic challenges, which today are a permanent element of the socio-economic environment, the role of enterprises and financial institutions in achieving sustainable development is emphasized (Hall et al., 2010; Baumann-Pauly et al., 2013; Ziolo et al., 2019b). Motivations to conduct socially responsible business activities in line with the SDGs can be assessed in terms of the benefits that the entity wants to achieve. Not every entity is guided by the adoption and implementation of the idea of sustainability solely by altruistic premises. Liston-Heyes and Certon (2007) claimed that few entities engage in social responsibility and sustainability for purely altruistic reasons. The practice shows that the entities try to achieve both their own benefit and have a positive impact on the company's environment by achieving SDGs. It is thought that long-term actions, including a strategy based on the idea of sustainability, should constitute a platform combining economic and social interests with a commitment to the environment. Sustainable entrepreneurship integrates the general concept of business entrepreneurship with SDGs, considering environmental, economic, and social issues (Allen & Malin, 2008; Hörisch, 2015), but it also integrates the concept of ESG risk with the risk associated with its business.

The economy needs to balance the expectations of all participants in social exchange. Therefore, both in the literature on the subject and in practice, the concept of stakeholders deserves attention. This concept is the basis for the implementation of the social, including ecological, responsibility of enterprises, including banks. Notably, according to this concept of the source of the value of an enterprise or a bank, it refers to both purely financial and non-financial aspects. From a financial point of view, value is analyzed from the perspective of the possibility of creating future profits or generating free cash flows. Non-financial factors are related to organizational aspects, marketing activities, and intangible resources, including human resources. Among the non-financial aspects are external and internal factors (Cetindamar & Husay, 2007) The external factors creating the value of a banking enterprise include, firstly, location, intensification of competition, infrastructure related to the economic system, legal regulations (especially in line with EU law and relating to security), economic environment and the market work, the shareholder factor (owners), and the perception of natural resources, including the policy toward sustainability. Internal factors relate to intellectual capital and the role of human capital, organizational culture, vision, mission, strategy, customer loyalty, products and services offered, innovation, quality, process uniqueness, compliance with CSR principles, and supporting R&D. For a long time, the literature on CSR has focused on companies' responsibility and their ethical principles (Carroll, 1979; Matten & Moon, 2008; Windsor, 2006; Wood, 1991). In addition, the impact on the environment through products, services, policies, and institutional strategies is important. The concept of stakeholders is much more broadly understood than the concept of shareholders, corporate responsibility, and applying it to all stakeholders, including society, and their perception of responsibility for sustainable development and the environment.

No clear answer about the motives for implementing sustainability in business is available in the literature, in particular for a banking enterprise. Unequivocal research on the direct influence of enterprises as clients on bank decisions in terms of policy and the perception of sustainability is also lacking. Several researchers have investigated the influence of corporate social responsibility (Carroll, 1999) and sustainability paradigms (Edwards, 2005; Moon, 2007; Pearce & Atkinson, 1993) on business and social relations. Indirect influence is discussed and, so far, it is widely identified with the concept of stakeholders. Themes for sustainability may come from at least the following directions:

- the responsible people, their views and feelings, or legal, financial, and operational assignments (Pellicano et al., 2014);
- influence: people who are, or might be, able to influence the ability of a company/organization to meet its goals, then goals in terms of sustainability (Leal Filho & Brandli, 2016);
- through proximity and longstanding relationships: the people with whom a company/organization interacts enforces behavior consistent with the SDGs (Zimmermann, 2019);
- through representation, which should be understood as a susceptibility to influence heads of a local community, trade union representatives, councilors, representatives of membership-based organizations, etc. (Krick et al., 2005).

Notably, after examining the impact and significance of the CSR concept, the integration of sustainability (as the next step) into corporate strategies (and in particular bank strategies) requires acknowledgment, in terms of impact and development, of social policies and procedures to reduce and control the potential negative impacts of corporate activities on stakeholder engagement and on an organization's social perception (Pellicano et al., 2014). It has also become the end to link the directional goals of the corporate strategies of banks. CSR is focused on the evolution of traditional corporate social responsibility into a set of actions oriented to a shared value creation and to the sustainability of corporate performance. Sustainability represents a global idea, concerning not only the efficient allocation of resources over time, but also their fair distribution between present and future generations (Boudreau et al., 2008; Ciegis et al., 2009).

In the sustainability debate, banking institutions are, conventionally, considered a passive player because their role is, generally, limited to obtaining financial liabilities, and distributing those liabilities in a neutral way. Here, then, a question should be asked as to whether this role is actually neutral? The actual motives and the role indicated in the literature are focused on the impact through banking policies and strategies, with two dominant directions: (1) social and environmental responsibility initiatives in internal bank operations and (2) integration of CSR in banking intermediation (social and environmental considerations in product design, credit policy, etc.) in stock market investments and other sources of financing investments (Pellicano et al., 2014; Zimmermann, 2019).

The literature indicates at least a few reasons for implementing the idea of sustainability in business: external coercion, economic conditions, competitive struggle, prices of raw materials and their derivatives (energy, raw materials), costs of waste disposal or storage, strategic premises of the enterprise, and moral and ethical premises (Bansal & Roth, 2000; Cassells & Lewis, 2011; Zaman, 2012; Ferenhof et al., 2014; Pellicano et al., 2014; Johnson & Schaltegger, 2016; Hughes & Troy, 2017). Although these are clear premises for an enterprise, for financial institutions, including banks, not all of them constitute the basis for building strategies and business models based on sustainability. In Table 5.1, the differences concerning the implementation of sustainability between enterprises and financial institutions, including banks, are presented synthetically. The illustration of the motivation considers the approach to ESG risk, i.e., environmental, social, and governance factors.

Table 5.1 Synthetic approach to the motivation to implement the principles of sustainability in enterprises and financial institutions

Categories of environmental motivations	*Categories for enterprises*	*Categories for financial institutions*
Environmental	competitive struggle based on the use of the environment; availability of raw materials; environmental awareness; strategic premises of the enterprise; efficient use of resources	strategic premises of the institution; pressure of stakeholders to support the environment; integrated environmental protection based on green sources of financing
Social	moral and ethical premises; relational and social motivations; social responsibility	moral and ethical premises; relational and social motivations; social responsibility; influence of public opinion
Governance	external coercion; economic conditions; competitive struggle; energy and raw material prices; costs of waste disposal or storage; strategic premises of the enterprise	external coercion; economic conditions; competitive struggle; strategic premises of the institution

Source Own elaboration

Bansal and Roth (2000) indicated a slightly different approach to the classification of categories of incentives related to the implementation of sustainability. The motivations are ethical, competitive, and relational. Ethical motivations result from the values represented in the company or financial institution, especially in the field of social responsibility for the environment, use of resources, and/or respect for human rights. Competitive motivations result from the adopted strategy, in which sustainability is a factor that allows for increasing income and shaping a competitive advantage. Competitive motivations include cost motivations (based on productivity, also called operational motivations, e.g., joining the offer with state interventionism in terms of sustainability) and motivations based on differentiation, which are the commercial incentives. Relational motivations are based on the pursuit of establishing the best possible relationships with various stakeholder groups in the institution. Research by

Henriques and Sadorsky (1999) suggested that institutions may be motivated to undertake ecological transformations due to stakeholder pressure (environmental pressure).

Russo and Fouts (1997) indicated that the companies most active in environmental issues, as customers of financial institutions, can, by creating a kind of environmental lobby, influence the policies, strategies, products, and services of financial institutions. Thus, both entities (enterprises and financial institutions) ensure a competitive advantage for each other. These studies were expanded and developed, which led the conclusion that it is commercial motivations that affect the image of a financial institution (González-Benito & González-Benit, 2005; Yip & Bocken, 2018; Kumar et al., 2016; Zimmermann, 2019).

5.3 The Roles of the Financial Sector in Sustainable Development

Financial institutions, in particular banks, can exercise considerable influence—in some cases, even control—over investment and management decisions, which could be brought into by enterprises for the benefit of the environment. Notably, the literature on the subject indicates the following interactions of financial institutions, including the interactions of banks, with the environment (Report, 1997):

1. As investors: supplying the investment and capital needed to achieve sustainable development;
2. As stimulating innovative activities: developing new financial products to encourage sustainable development, e.g., in energy efficiency, in technological changes supporting recycled products;
3. As valuers: pricing risks and estimating returns for companies, projects, and others; assessing the impact of ESG risk and estimating the possibility of its elimination;
4. As powerful stakeholders: as shareholders, equity providers, guarantors, and lenders, they can exercise considerable influence on the management of companies;
5. As polluters: while not dirty industries, financial institutions consume considerable resources;
6. As victims of environmental change: along with changes in mentality, they need to implement measures to reduce ESG factors and implement a wider range of products supporting sustainability in interactions with customers changes, e.g., from climate change.

As the European Commission identified three main actions for sustainable economic growth, actions that are supported by financial inscriptions, including banks, must focus on capital flows toward environmentally and socially friendly investments, integrating sustainability in risk management, promoting transparency, and increasing the incentives for the implementation of long-term value-building strategies (ONZ, 2015). These activities coincide with activities aimed at eliminating ESG risks. This situation led to the development of activities and the formulation of goals for financial institutions, including banks, which should be expressed in their operational strategies considering sustainable economic growth. In particular, the actions of financial institutions and banks should lead to the following (ONZ, 2015):

- private capital mobilization to reduce the estimated financial gap in connection with the need to transform the economy in line with the paradigm of sustainable development;
- stimulating technological development, including innovations conducive to sustainability and elimination of ESG risks;
- disseminating knowledge, methodological frameworks, and good practices related to sustainable development and promoting change in the field of corporate governance based on sustainability.

Figure 5.1 shows the role of financial institutions, including banks, in creating a green order, which is the basis of sustainability. Therefore, the role of financial institutions in supporting sustainable development is providing financial knowledge, especially capital. This is a classic approach to the role of intuitions in the economy, but their role can be perceived through the prism of factors limiting ESG risk. The literature on the subject indicates the following modern environmental role of financial institutions, including banks (Dziawgo, 2010; Janicka, 2016):

- Engagement of financial institutions in environmental protection: under the influence of pressure from stakeholders and along with the increasing environmental awareness of the people who manage these institutions, actions are taken (including a strategy being built) to support the idea of sustainability.
- Motivation to undertake environmental measures: the approach to environmental issues in financial institutions is changing, especially

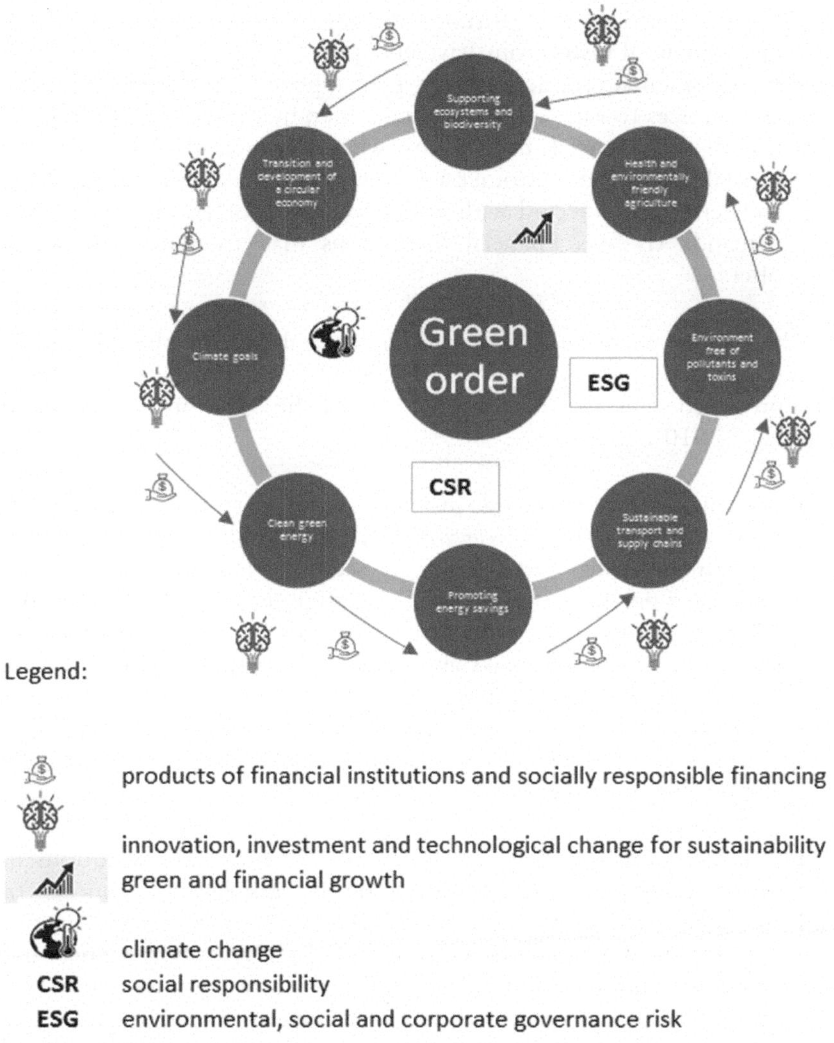

Fig. 5.1 The role of financial institutions, in creating a green order as a key factor of sustainability (*Source* Own elaboration)

because the views and attitudes of those who have a decisive impact upon financial institutions have changed.

- Environmental products and services: adjusting the offer of products and services toward environmentally friendly products and services.
- Clients of financial institutions who are interested in environmental protection: customer attitudes generate demand for products and services connected with environmental protection; hence, their demands vis-à-vis financial institutions may translate into actual offers.

The role of financial institutions in legal alignment should not be forgotten. With respect to financial markets, the European Commission proposed inter alia measures to support the SDGs in Europe 2020 (Strategy, 2010):

- designing new financing instrument, in particular in cooperation with the EIB (*European Investment Bank)*/EIF (European Investment Fund) and the private sector, responding to hitherto-unfulfilled needs by businesses. This design for countries outside the EU may be associated with dedicated financial instruments that are supported by central banks and agree with them as part of strategic activities for the environment;
- making an efficient venture capital market a reality, thereby facilitating direct business access to capital markets and environmentally responsible investments.

In the literature on the subject, the role of financial institutions in sustainable development is expanded. In this respect, the concept of climate finance (defined by the UNFCCC) is narrow and refers to finance that aims at reducing emissions, enhancing sinks of greenhouse gases, and reducing vulnerability of, and maintaining and increasing the resilience of, human and ecological systems to negative climate change impacts. The broader role of financial institutions supporting SDGs and the idea of sustainability should be understood as financing activities by banking and non-banking financial institutions with the aim to reduce greenhouse gas emissions and increase the resilience of the society to negative climate change impacts while considering other sustainable development goals such as economic growth, job creation, and gender equality. This understanding considers both classic financing and a set of measures to mitigate the effects of ESG risks (Park & Kim, 2020; UNEP, 2016).

To summarize, behind the practical solutions, the current role of financial institutions, including banks, in supporting sustainable development focuses on the following (Forstater et al., 2016):

1. Mobilizing finance for specific sustainable development priorities and to mainstream sustainable development factors across financial decision-making to enable the achievement of SDGs;
2. Considering mainstream sustainable development factors across financial decision-making.

The role of financial institutions (Osborn et al., 2015; Forstater et al., 2016; Filipiak & Dylewski, 2020) is to:

1. Mobilize capital for the inclusion of underserved groups (e.g., small and medium enterprises) and to implement underserved projects with high environmental innovation;
2. Conduct activities to create products enabling raising capital for sustainable infrastructure (e.g., energy, housing, transport, and urban design) and financing critical areas of innovation (e.g., agriculture, mobility, and power);
3. Create with sustainability as a mainstream for financial institutions;
4. Create a common platform for assessing the ability to meet debt obligations (when raising capital) to ensure the integrity of the impact of direct factors used in creditworthiness assessments (e.g., past creditworthiness assessment procedures) and indirect (e.g., tax, corruption, and human rights), which extends to integrating environmental and social (E&S) factors into risk management (e.g., climate disruption, water stress);
5. Enforce universal ESG reporting, which means enabling environmental reporting and ESG into the responsibilities and reporting of market actors to guide their decision-making.

As sustainability is of particular importance due to the problems of preserving natural resources and quality of life and limiting the effects of ESG risk, it must be assumed that actions to achieve the SDGs are the basic condition for the continued existence and evolution of humanity, climate protection, and the ecosystem. Hence, sustainability should not be determined only by the economic situation: it should be determined by the actual involvement of leading institutions in the creation of legal, economic, and educational mechanisms, including financial institutions. The aim of this commitment is to stimulate the development of pro-ecological production of goods and conscious consumer attitudes in accordance with the principle of sustainable development through the law, economic incentives, and financial products. The scope of activity of financial institutions, including banks, as an important element of the indicated impact system includes, in particular, the following projects and initiatives: eco-economy, environmental reporting, ad hoc environmental initiatives, and offering and selling financial products and services. Table 5.2 presents the classification of actions taken according to the direct or indirect economic activity criteria of a financial institution.

Direct involvement in the sustainability and implementation of SDGs requires institutions, including banks, to modify both their operating strategies and existing business models. In addition, a decision is required as to whether financial institutions, including banks, will directly support SDGs by sharing profit or sponsoring and promoting environmental activities. It is difficult to indicate whether a direct role or an indirect influence will benefit to the achievement of the SDGs due to the variety of recipients of this impact (clients and other stakeholders) who can adapt to coercion or increasing environmental awareness.

As previously indicated, actions for sustainability by incorporating this idea into the business models of financial institutions will require them, firstly, to redefine their revenue sources and financing sources and, secondly, to define the structure of revenue diversification, financing sources, operating markets, and key clients. This situation may lead to a partial loss of customers, markets, and sources of revenues, as well as to the debugging of a new customer base, entering new markets (especially innovative ones), and building relationships with the public sector that support the achievement of SDGs.

Table 5.2 Classification of actions taken according to the criterion of direct or indirect economic activity of a financial institution taking into account ESG risks

Environmental	carrying out the practices of purchasing materials and devices, taking into account the environmental impact as one of the important criteria	indirect economic activity
	conducting rational waste management in financial institutions, especially limiting their generation, selective collection and recycling or disposal	indirect economic activity
	limiting the use of raw materials and natural resources through rationalization of consumption of water, electricity, heat and fuels	indirect economic activity
	offering products and services to finance pro-ecological business ventures and raising funds (products used to raise capital for pro-ecological initiatives) in order to finance activities related to environmental protection	direct economic activity
	integrating ecological risk into the process of financing business ventures of clients of financial institutions	direct economic activity
Social responsibility	popularizing knowledge about pro-ecological behavior among employees and clients, raising awareness and shaping attitudes toward sustainability	indirect economic activity
	cooperation with institutions and organizations whose goals are focused on sustainability and CSR	indirect economic activity
	investor relations toward sustainability	indirect economic activity
	public and society relations toward sustainability	indirect economic activity
	compliance with CSR standards, in particular with regard to working conditions and compliance with employee rights and health and safety rules	direct economic activity

(continued)

Table 5.2 (continued)

	reporting on social responsibility (in particular ensuring GRI standards), their popularization and informing the society	direct economic activity
	promoting and sponsoring activities for sustainable development in the environment	direct economic activity
Corporate governance	respecting applicable law in the field of environmental protection and taking into account its requirements in the internal regulations of a given financial institution	indirect economic activity
	introducing reporting on social responsibility standards, introducing pro-ecological accounting standards	direct economic activity
	developing, implementing and modifying business models and action strategies based on achieving DSG's goals	direct economic activity
	direct promotion of sustainability through profit sharing in order to implement projects related to environmental protection and CSR	direct economic activity

Source Own elaboration

5.4 Role of Central Banks in Enhancing Sustainability

After the recent financial crisis, central banks have paid special attention to the issues of financial system stability. Increasing attention is also paid to the idea of a sustainable financial system (Ziolo et al., 2019a, 2019b).Central banks realize that a green transformation toward sustainable is inevitable. Therefore, taking into account the demand for capital aimed at financing activities resulting from the green transformation and the implementation of the SDGs' objectives, in particular for financing projects, innovations, and investments toward low-carbon development, the financial sector will have to play a central role in allocating resources and stop financing activities that harm the environment (Volz, 2017).

More and more voices are being heard about the wider central bank mandate in affecting sustainability. The literature describes the environmental role of central banks. Along with the changes after the last financial crisis, the thesis was put forward that the renewed emphasis on the role of central banks in safeguarding financial stability can be directly linked to the case for responding to environmental and sustainability challenges (Volz, 2017). Discussions of practitioners, in particular between valuable banks, on climate change and central banking led to the formulation of two strong arguments:

> (1) climate change and society's response to it create risks that threaten the stability of the financial system, which it is central banks' duty to monitor and protect; (2) as public institutions responsible for regulating finance and creating public money, central banks should contribute toward closing the green investment gap. (Macquarie, 2018)

The newly established Network for Greening the Financial System (NGFS) plays an important role in strengthening the sustainability and achieving SDGs. Thanks to this network, central banks gained the possibility of quick exchange of experience and ideas between central banks and supervisors in three areas: microprudential supervision, macroprudential analysis and measures to scale up green finance (Breitenfellner et al., 2019). The latest report by the NGFS (2019) provides the following recommendations (NGFS, 2019):

1. integrating climate-related risks into financial stability monitoring and microsupervision,
2. integrating sustainability factors into their own portfolio management,
3. bridging data gaps, and
4. building awareness and intellectual capacity and encouraging technical assistance and knowledge sharing.

NGFS, at the same time formulating recommendations, calls for actions for sustainability.

In summary, if the SDGs are not directly part of a central bank's explicit mandate, the instrumental incorporation of sustainability factors may be relevant to achieve price stability or safeguard financial stability. Table 5.3 presents three broad types of arguments justifying why central banks should respond to environmental and sustainability challenges.

Table 5.3 Three broad types of arguments justifying necessity respond to environmental and sustainability challenges

Types of arguments	*Explanation of arguments*	*Influence on the role of the central bank*
The financial and macroeconomic risk argument	Climate change and environmental damages may have very direct consequences for price stability through their impact on food and energy prices Floods and droughts may affect agricultural output, which in turn influences food prices The need for climate change mitigation impacts patterns of energy production and energy prices	Monetary policy has to take into account changes to food and energy prices, as these changes can have indirect and second-round effects on core inflation and this fact must be taken into account in central banks' long-term inflation outlook analysis
	There may be risks to financial stability from environmental damage and climate change	Central banks need to move away from narrow price stability-focused mandates and take charge again of financial stability and their role should focus in mitigating systemic risk by identifying system-wide vulnerabilities and using their panoramic view of the financial system to connect the sensitive dots
	Environmental risk and risk from climate change a significant systemic risk for the financial sector	The role of the central bank is to stimulate changes in the rules that guide and govern the operation of financial markets need to evolve to address the systemic risk; banks and other financial institutions to price in these risks

(continued)

Table 5.3 (continued)

Types of arguments	*Explanation of arguments*	*Influence on the role of the central bank*
	Risks of pervasive disruptions to individual businesses and entire industries or even economies stemming from climate change may be significant	
	Significant impact of "transitional risk." High carbon and resource intensity have a long-term impact on the viability of individual businesses and the competitiveness of economies at large, which implies also risks for macroeconomic sustainability and stability	If the re-pricing of risk happens in a gradual manner, the risks to financial stability will be limited, but financial authorities may have a role in safeguarding that these risks are addressed
	Disclosure is central to addressing climate and environmental risk and inadequate information on risk exposures can lead to a mispricing of assets and/or misallocation of investment and can potentially give rise to concerns about financial stability	These risks need to be considered in central banks' financial stability and macroprudential policy frameworks
The Market Failure Argument	The occurrence of negative effects of credit allocation by banks is related to the allocation of loans in a suboptimal manner from a societal perspective, and the effect of the decision is with too much being allocated to carbon-intensive, polluting activities	The central bank and the government may resort to a best policy and government can mandate the central bank or financial supervisor to address negative environmental externalities by using the tools they have at their disposal

(continued)

Table 5.3 (continued)

Types of arguments	*Explanation of arguments*	*Influence on the role of the central bank*
	Environmental regulation and carbon pricing should be the preferred policy tools to correct this market failure and prevent or disincentivize such investments and the central bank may use its powers to affect credit creation and allocation	
	Market failure is missing markets or segments of market related to sustainability	Central banks take action to support the build-up of missing market segments to promote green finance, a ensure that traditional tools of monetary policy are made environmentally consistent
Central Banks as Credible and Powerful Actors in Developing Countries	The strong institutional standing of central banks have in the policy frameworks of countries or communities of countries; different levels of development of individual countries (highly developed countries or developing countries have completely different attitudes toward sustainability) determine the role of central banks	Central banks their command over the banking sector, they can effectively exert influence over private investment decisions. Moreover, central banks' financial market expertise and their transnational networks can help them to promote "best sustainable practice" reforms in the financial sector
	Central banks as promoters of sustainable development	The developmental role of central banks
	The mandate and accountability of central banks, potential conflicts issue between goals of central banks and their institutional independence	Strengthening central banks' credibility as institutions supporting sustainability, implementation of SDG's goals and creators of the policy for sustainability

Source Own elaboration based on Epstein (2009), Macklem (2011), Leaton et al. (2013), Carney (2015), Kiernan (2014), Volz (2017)

Central banks, depending on the level of socio-economic development of the country, try to be active in their response to the calls to green the financial system and its growth. They use a variety of instruments to help green the financial system and respond to systemic environmental risks. For this purpose, they often adapt the tools used so far. What stands for recalibrated tools with a green focus. Central banks also use new instruments in the area of macroprudential regulation (Dikau & Ryan-Collins, 2017). Table 5.4 describes the recommended taxonomy of green instruments, taking into account three different areas of central bank activity.

Sustainable development and the transition to a low-carbon, more resource-efficient, circular economy are key elements in ensuring the long-term competitiveness of the EU economy. Financing supports the economy by providing funds for economic activity, with the aim of creating jobs and stimulating economic growth. Investment decisions are usually made on the basis of a number of factors; therefore, they are seen as a significant role of central banks and are included in the European and global policy for action to achieve the SDGs. Sustainable funding and central bank policy generally refer to integrating environmental and social aspects into investment decision-making, integrating ESG risk into the assessment, leading to increased investment for sustainable development. A special place is assigned to social issues in the EU zone. European countries have raised social issues in the role of central banks. They indicate that in financing sustainable development, social issues can relate to issues of inequality, social inclusion, labor relations, and investment in human capital and communities. Environmental and social issues are often interrelated. The existence of environmental risks can exacerbate existing systems of inequality. Good practices, training, and the implementation of guidelines (as part of the policy by central banks) allow for the development of a new management model for structures, relationships with employees, and remuneration of management, playing a key role in ensuring that social and environmental issues are considered in the decision-making process (UE, 2018).

The case for central banks to pursue sustainability objectives goes beyond their traditional core mandates. Central banks must interact with financial institutions to influence their business models, but they must also be involved in sustainable development policy, which is the domain of governments. Having appropriate instruments at their disposal (Table

Table 5.4 Taxonomy of green instruments of central bank activity

Areas of central bank activity	*Type of*	*Characteristic of instruments*
Green credit allocation instruments that have the objective of allocating credit to green sectors	Targeted refinancing lines	Targeted on sustainability goals refinancing lines offer refinance for commercial banks at preferred terms for specified asset classes The problem with this instrument is the potential loss of control over the cash base and the fact that differential interest rates can lead to distortions in the financial system in the long run. This instrument gives a certain flexibility of decision even though credit is extended at preferred terms but commercial banks still decide which projects are financed
	Differential reserve requirements	A green reserve requirement policy would allow banks to hold fewer reserves—which bear zero or low interest—against green loans. This resulted in the necessity to align the profitability of their lending activities with the policy target of greening finance and growth
	Minimum and maximum credit quotas	The tools are priorities sector lending programs, with the central bank determining minimum credit quotas that have the objective of greening growth by requiring commercial banks to lend a specific percentage of their overall lending to green projects

(continued)

Table 5.4 (continued)

Areas of central bank activity	*Type of*	*Characteristic of instruments*
	Central bank assistance to development banks	Development banks to help finance the green transition of the economy and to overcome the absence of long-term patient capital in the sustainability sector. These banks, having sufficient capital from governments, could broaden their mission with central bank financing their lending capacity and legitimacy could be significantly increased
Green Macroprudential Policy Instruments	Stress testing	Stress testing is used to evaluate and calibrate green macroprudential policy instruments. These activities are aimed at determining stress test various financial instability scenarios and enable them identified vulnerabilities, capital buffers, risk weights, and caps could then be set accordingly to mitigate systemic risk and reduce the threat to financial instability
	Counter-cyclical capital buffers	Counter-cyclical capital buffers may be used to mitigate potentially adverse effects related to tightening environmental standards. This tightening causes a demand for capital, and lending to adjustment investments is growing

(continued)

Table 5.4 (continued)

Areas of central bank activity	*Type of*	*Characteristic of instruments*
	Capital instruments: (1). Leverage ratio by sector (2). Loan-to-value (LTV)/loan-to-income (LTI) caps (3). Large exposure restrictions (by counterparty, sector, geographic)	Assigning assets infringing sustainability higher risk weights to take into account the fact that future environmental policies might reduce their value Limitation could be used to limit lending (financing) are companies or sectors whose business models are not based on sustainability activities likely to become unprofitable in the near future, given either new regulations or innovations in green technology This instrument would allow a certain amount of fine-tuning of lending restrictions; it also allows to limit exposure to carbon-intensive assets
	Identification of systemically important financial institutions (SIFIs) and capital surcharges	Use of exposure to carbon-intensive as criteria used to identify SIFIs
Other Green Central Banking Initiatives and Activities	Green finance guidelines	Central banks are in the position to develop green banking guidelines, they may be mandatory or voluntary
	Active development of green bond markets	Central banks conduct activities to support the development of financial markets and especially of bond markets. In particular, they may encourage the issuance of green bonds as a means of financing green projects and other governmental agencies can issue green bond guidelines and define criteria for environmentally friendly projects

(continued)

Table 5.4 (continued)

Areas of central bank activity	*Type of*	*Characteristic of instruments*
	Research on green growth and finance	Central banks support research for green finance and growth. The collected data and research have playing a central role in conducting further research in green growth and sustainability
	Offering capacity building workshop for bankers	Workshops and seminars appear in the offer of central banks for commercial banks on environmental risk assessment and green finance
	Participation in the International Finance Corporation's Sustainable Banking Network	Access and active participation in the network, exchange of experiences, promotes sustainable finance

Source Own elaboration based on Werner (2003), Gray and Talbot (2007), Mazzucato and Penna (2015), Schoenmaker and Tilburg (2016), Stern (2016), UNEP (2016), Dikau and Ryan-Collins (2017)

5.4), they can support government policy with their instruments by influencing commercial banks. An important aspect is the use of instruments of influence through which, in parallel with environmental objectives, risk (negative impact on the stability of the financial system) and ESG risk are eliminated. In particular, it is about ensuring a balance between environmental and social issues.

5.5 Green Bank Policies in the Corporate Sector

The literature on the green bank policies indicates that financial and political conditions may be unique to each country, but there are certain core elements to any national green policy in the bank development process (CGC, 2017). These elements include development of national interest and being a local champion, a market analysis across technologies and sectors, engagement of key stakeholders, consideration of and adaptation to the local enabling environment and policy framework, and recruitment of public and private sources of capital. Different strategies suit the differing institutional landscapes of individual nations.

Institutions of the financial system, and in particular banks, based on cooperation with the central bank, are focusing on environmental protection and achieving the SDGs. Their key role in this regard is based on three-way action. The first is to work toward ethicality and the second is to act for socially responsible banking as an important role in their corporate social responsibility (Tara et al., 2015). Considering the implementation of the SDGs, financial institutions, including banks, operate in such a way as to be environment-friendly and ethically correct, and conduct business in accordance with the idea of sustainability (Jha & Bhome, 2013; Mishra, 2013; Biswas, 2011; Tara et al., 2015).

The policy of banks in the corporate sector is based on the selection of assumptions from among two directions that are key to determining business models by banks:

- implications for business models resulting from the assumptions of the SDGs' business models;
- business models resulting from the essence of social banking.

The first direction results from the banks' position in the existing financial ecosystem, which is framed by financial regulation, shareholder expectations, and the rules of the game, like good practices. Clients and stakeholders are increasingly turning their attention to banks that are responsible and interested in protecting the environment (Rai et al., 2019). They are more likely to prefer banking institutions that involve them in decision-making through co-creation and considering green value. This decision approach, taking the aspect of the environment into account, builds a sense in consumers that this bank is concerned about the environment in real terms, which boosts their green loyalty (Wu & Cheng, 2018). Conversely, it is necessary to point to the financial regulations that result from the involvement of governments and communities in the implementation of the SDGs (membership in organizations, associations, adoption of resolutions, and creation of internal law). This situation means that governments and central banks are also creating tools and instruments that will favor environmental issues (as discussed in

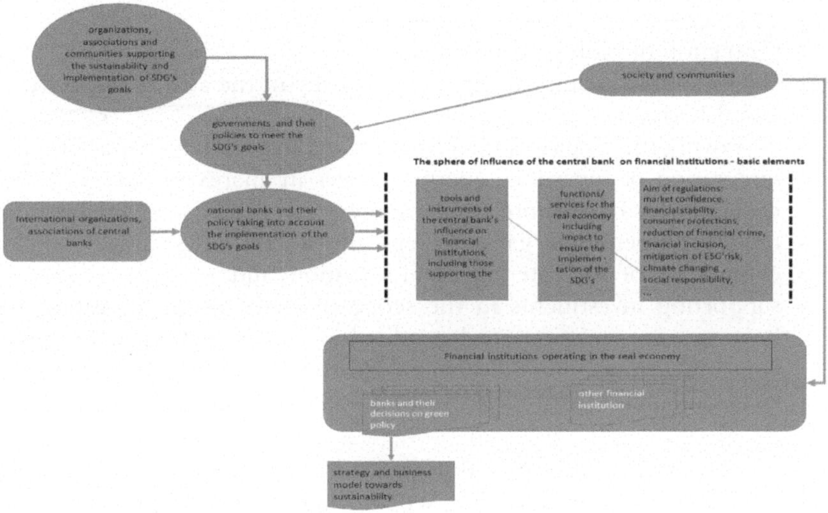

Fig. 5.2 The mutual interactions that influence the green policy of banks and their business models (*Source* Own elaborations)

Sect. 5.3). Mainly governments and national banks are required to facilitate the best possible financing and financial services consistent with the sustainability supply to the real economy and to society, without endangering the stability and liquidity of the financial sector and society's trust in financial markets (Kreibiehl & Patel, 2014). Figure 5.2 shows the mutual interactions that influence the green policy of banks and their business models when considering sustainability.

Banks, when implementing their strategies and business models, base them, as previously indicated, on the environmental strategies and policies of governments. They implement their own beliefs regarding the idea of sustainability, and they are part of the governmental guideline and support instruments. The most common areas of supporting sustainability introduced into banks' business models are the following[1]:

[1] https://ec.europa.eu/info/strategy/priorities-2019-2024/european-green-deal/actions-being-taken-eu/eu-biodiversity-strategy-2030_pl.
https://ec.europa.eu/commission/presscorner/detail/pl/fs_19_6727.
https://ec.europa.eu/commission/presscorner/detail/en/fs_19_6723.

- a low-carbon and resource-efficient economy;
- protection of biological diversity;
- changes introducing the circular economy in the area of agriculture and fisheries;
- improving the quality of interconnections and working toward better integration of renewable energy sources in energy systems and the development of offshore wind energy;
- reducing greenhouse gas emissions in the transport sector;
- elimination of air, water, and soil pollution; and
- supporting investments in the scope of enabling the transition to low- or zero-emission steel production and reduction in resource consumption in the following industries: clothing, construction, electronics, and plastics.

This will have a future and significant influence on banks' business strategies through the EC Action Plan on financing sustainable growth, which formulates the following goals: channeling capital flows toward sustainable investments to achieve sustainable growth, financial risk management resulting from climate change, environmental degradation and social problems, and promoting transparency and long-term sustainability in financial and business activities (EC, 2018).

The second option for determining business models by banks toward green banks implementing the sustainability policy is social banking. The idea of social banks indicates that social and economic sustainability are considered when making financial decisions (Benedikter, 2011). Building business models that take into account the assumptions of the social banking concept forces the implementation of the triple principle of profit–people–planet. The concepts of responsibility (individualism and ethics of decisions made), transparency (transparency of transactions, full insight into the course of transactions, and reporting), and sustainability (the aim to develop society, the environment and individuals,

https://www.osw.waw.pl/pl/publikacje/analizy/2020-07-10/unijna-strategia-wodorowa.

https://ec.europa.eu/commission/presscorner/detail/pl/fs_19_6724.

https://ec.europa.eu/commission/presscorner/detail/pl/fs_19_672.

https://ec.europa.eu/commission/presscorner/detail/pl/fs_19_6725.

https://ec.europa.eu/commission/presscorner/detail/pl/fs_19_6729.

and making investments in small projects that are socially and environmentally friendly) should also be included in the construction of banks' business models (Benedikter, 2011). When building business strategies, banks should follow the principles derived from their double bottom line (Becchetti et al., 2011; San José et al., 2011). This means that profit-making is a means for achieving economic sustainability, assuming no speculating on financial markets and focusing on financing the real economy (Cornée et al., 2016).

Social banking is a path in which banking models are built considering environmentalism and society issues. The two paths indicated of building green policy included in banks' business models only show the components for which instruments and tools, and even entire programs, should be selected. For this selection to be correct, precision of action resulting from the adopted strategy is necessary for the business model to be effective.

In practice, many banks, especially private sector commercial banks, have started modifying their business models, introducing greening to their operations by integrating environmental and climate change risks into their strategies and risk management systems, rolling out green financial products, and promoting sustainability and social responsibility (Su et al., 2020). Precision is essential in banks' strategic assumptions regarding sustainability, as well as in introducing individual elements of the sustainability concept to business models. Therefore, despite the conceptual discussion and the lack of uniform terminology, the individual components of the sustainability concept used to determine the type of impact on the bank's business models can be highlighted (Park & Kim, 2020).

Figure 5.3 presents the terminology that allows defining the components of banks' business models. The presented approach allows for the formulation of green bank policy based on the constituent ideas of sustainability.

The purposes of the green policy of banks addressed to the companies sector are as follows (Weber, 2016):

- to promote green tools used to achieve green growth in the corporate sector;
- to promote models, ideas, and good practices related to ESG risk, sustainability, and the green economy, ensuring transparency of transactions and business, and social responsibility;

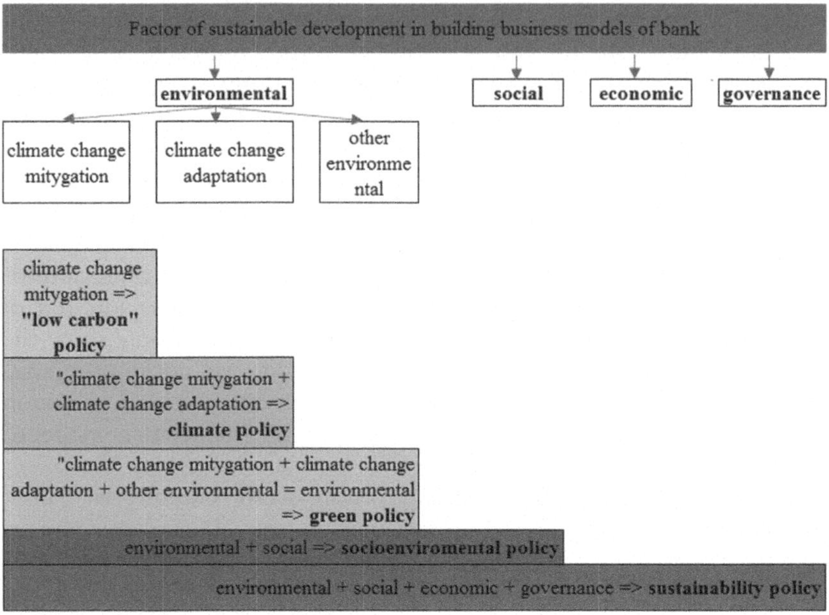

Fig. 5.3 Terminology that allows to define the components of banks' business models taking into account the green policy and the idea of sustainability (*Source* own elaborations based on: UNEP [2016]; Park & Kim [2020]

- to promote green financial instruments as a strategy; support the economy to grow in a green, low-carbon, and recycled model through business innovation; manage environmental and social (E&S) risks; improve own E&S performances; and, in doing so, optimize products' structure, improve services, and contribute to the transformation of the economic growth pattern;
- to indicate ESG risk management systems, and, in particular, to indicate how to effectively identify, assess, monitor, control, or mitigate ESG risks in business operations (such as energy consumption, pollution, land, health, safety, resettlement, ecosystem protection, climate change, etc.); and
- to promote green instruments raising capital for green investments and innovations.

The green banking policies require a fundamental change in the planning process with adequate consideration of the economy, business, finance, society, and the banks' profit (Tara et al., 2015). The construction of banks' business models should therefore not only include the factors indicated in Fig. 5.3, such as environmental, social, governance, and economic factors: it should also be based on the proportions that currently arise from the challenges of changing sustainability and environmental conditions. Banks should keep pace with their corporate policies to ensure they play a servant role in the face of the sustainability challenges faced by enterprises.

5.6 Conclusions

Banks and other financial institutions usually have no (or a negligible) impact on the natural environment, but they understand that they have to make decisions to be more proactive in this field. However, to achieve a more sustainable form of bank development, a more holistic perspective is needed that allows the consideration of all the resources upon which their business relies. Therefore, it is necessary to know and clearly define the motives of banks in the direction of sustainability. Banks, moving toward sustainability, formulate their role in the process of achieving the SDGs goals. Commercial banks see the central bank as a partner in the implementation of SDGs goals, especially if the central bank contributes to sustainability.

Banks have learned that attendance to environmental issues allows them to uncover latent and potentially harmful risks, save money and simultaneously improve their social responsibility reputation. This is a situation where private sustainability is also associated with improving social sustainability. The policies and activities of banks will have to respond to the challenges of sustainability to be able to build and finance environmentally friendly activities in the business sector and to change attitudes through cooperation with the bank to a more environmental-directed activities. The pandemic also shows the new role of banks and new risks, especially non-financial ones. Therefore, banks must take into account in their activities and in their policy toward sustainability a number of as yet unfinished non-financial factors, and those considered as ESG risks may have a stronger or different expression. The pandemic situation significantly increases the risks in banks, and these risks must be incorporated into modified business models.

References

Allen, J. C., & Malin, S. (2008). Green entrepreneurship: A method for managing natural resources? *Society & Natural Resources: An International Journal, 21*(9), 828–844.

Bansal, P., & Roth, K. (2000). Why companies go green: A model of ecological responsiveness. *Academy of Management Journal, 43*(4), 717–736.

Baumann-Pauly, D., Wickert, C., Spence, L. J., & Scherer, A. G. (2013). Organizing corporate social responsibility in small and large firms: Size matters. *Journal of Business Ethics, 115*(4), 693–705.

Becchetti, L., Garcia, M., & Trovato, G. (2011). Credit rationing and credit view: Empirical evidence from loan data. *Journal of Money, Credit and Banking, 43*(6), 1217–1245.

Benedikter, R. (2011). *European answers to the financial crisis: Social banking and social finance*. Spice Digest.

Biswas, N. (2011). Sustainable green banking approach, the need of the hour. *Business Spectrum, 1*(1), 32–38.

Boudreau, M. C., Chen, A., & Huber, M. (2008). Green IS: Building sustainable business practices. *Information Systems: A Global Text*, 1–17. https://www.researchgate.net/publication/242112741_Green_IS_Building_Sustainable_Business_Practices.

Breitenfellner, A., Pointner, W., & Schuberth, H. (2019). The potential contribution of central banks to green finance. *Vierteljahrshefte Zur Wirtschaftsforschung, 88*, 55–72.

Carney, M. (2015, September 29). *Breaking the tragedy of the horizon—Climate change and financial stability*. Speech given at Lloyd's of London. www.bankofengland.co.uk/publications/Pages/speeches/2015/844.aspx.

Carroll, A. B. (1979). A three-dimensional conceptual model of corporate social performance. *Academy of Management Review, 4*, 497–505.

Carroll, A. B. (1999). Corporate social responsibility evolution of a definitional construct. *Business & Society, 38*(3), 268–295.

Cassells, S., & Lewis, K. (2011). SMEs and environmental responsibility: Do actions reflect attitudes? *Corporate Social Responsibility and Environmental Management, 18*, 186–199.

Cetindamar, D., & Husay, K. (2007). Corporate social responsibility practices and environmentally responsible behaviour: The case of the United Nations global compact. *Journal of Business Ethics, 76*(2), 163–176.

CGC (Coalition for Green Capital). (2007, July). *National green banks in developing countries: Scaling up private finance to achieve Paris*. Climate goals coalition for green capital with support from the Hewlett Foundation.

Ciegis, R., Ramanauskiene, J., & Martinkus, B. (2009). The concept of sustainable development and its use for sustainability scenarios. *Inzinerine Ekonomika-Engineering Economics, 2*(62), 28–37.

Cornée, S., Kalmi, P., & Szafarz, A. (2016). Selectivity and transparency in social banking: Evidence from Europe. *Journal of Economic Issues, 50*(2), 494–502.

Dikau, S., Ryan-Collins, J. (2017). *Green central banking in emerging market an developing country economies*. New Economics Foundation

Dziawgo, L. (2010). *Zielony rynek finansowy*. Ekologiczna ewolucja rynku finansowego.

EC Action Plan on financing sustainable growth, UE Commision. (2018). https://www.unpri.org/download?ac=5173.

Edwards, A. R. (2005). *The sustainability revolution: Portrait of a paradigm shift*. New Society Pub.

Epstein, G. A. (2009). *Post-war experiences with developmental central banks: The good, the bad and the hopeful* (G-24 Discussion Paper No. 54). Geneva: UNCTAD.

Ferenhof, H. A., Vignochi, L., Selig, P. M., Lezana, Á. G. R., & Campos, L. M. S. (2014). Environmental management systems in small and medium-sized enterprises: An analysis and systematic review. *Journal of Cleaner Production, 74*, 44–53.

Filipiak, B. Z., & Dylewski, M. (2020). Sustainable financial reporting in the context of ensuring sustainability of financial systems. In M. Ziolo (Ed.), *Finance and sustainable development* (pp. 235–267). Routledge.

Forstater, M., Halle, M., & Zade, S. (2016). *Green finance for developing countries*. Needs, concerns and innovations, United Nations Environment Programme, Washington, D.C.

González-Benito, J., & González-Benito, Ó. (2005). A study of the motivations for the environmental trans-formation of companies. *Industrial Marketing Management, 34*(7), 462–475.

Gray, S., & Talbot, N. (2007). *Developing financial markets*. Bank of England.

Hall, J. K., Daneke, G. A., & Lenox, M. J. (2010). Sustainable development and entrepreneurship: Past contributions and future directions. *Journal of Business Venturing, 25*(5), 439–448.

Henriques, I., & Sadorsky, P. (1999). The relationship between environmental commitment and manage-rial perceptions of stakeholder importance. *Academy of Management Journal, 42*(1), 87–99.

Hörisch, J. (2015). The role of sustainable entrepreneurship in sustainability transitions: A conceptual synthesis against the background of the multi-level perspective. *Administrative Sciences, 5*, 286–300.

Hughes, R. C., & Troy, M. (2017). Motivations of one small business community to implement environmental sustainability as a business practice. *Journal of Environmental Sustainability, 5*(1), Article 2, 1–8, Available at: https://scholarworks.rit.edu/jes/vol5/iss1/2.

Janicka, M. (2016). Financial markets and the challenges of sustainable growth, comparative economic research. *Central and Eastern Europe, 19*(2), 27–41.

Jha, N., & Bhome, S. (2013). A study of green banking trends in India. *International Monthly Referred Journal of Research in Management and Technology, 2,* 127–132.

Johnson, M. P., & Schaltegger, S. (2016). Two decades of sustainability management tools for SMEs: How far havewe come? *Journal of Small Business Management, 54,* 481–505.

Kiernan, P. (2014, September 23). *Risky business*. The Economist Intelligence Unit. www.economistinsights.com/energy/opinion/risky-business.

Kreibiehl S., & Patel S. (2014, March). *Delivering the green economy through financial policy*. Frankfurt School of Finance & Management—UNEP Collaborating Centre for Climate & Sustainable Energy Finance.

Krick, T. et al. (2005). *From words to action the stakeholder engagement manual Vol. 2: The practitioner's handbook on stakeholder engagement*. Available at: https://www.accountability.org/images/content/2/0/208.pdf.

Kumar, V., Rahman, Z., & Kazmi, A. A. (2016). Assessing the influence of stakeholders on sustainability marketing strategy of Indian companies. *SAGE Open, 6*(3), 1–19.

Leal Filho, W., & Brandli, L. (2016). Engaging stakeholders for sustainable development. In W. Leal Filho & L. Brandli (Eds.), *Engaging stakeholders in education for sustainable development at university level*, World Sustainability Series (pp. 335–342).

Leaton, J., Ranger, N., Ward, B., Sussams, L., & Brown, M. (2013). *Unburnable carbon 2013: Wasted capital and stranded assets*. Carbon Tracker & The Grantham Research Institute, London School of Economics and Political Science.

Liston-Heyes, C., & Ceton, G. (2007). Corporate social performance and politics, do liberals do More? *The Journal of Corporate Citizenship, 25,* 95–108.

Macklem, T. (2011, June 6). *Mitigating systemic risk and the role of central banks*. Remarks to "Conférence de Montréal", Montreal, Quebec. www.bis.org/review/r110610c.pdf.

Macquarie, R. (2018). *A green bank of England positive money: Partners for a new economy*.

Matten, D., & Moon, J. (2008). "Implicit" and "explicit" CSR: A conceptual framework for a comparative understanding of corporate social responsibility. *Academy of Management Review, 33,* 404–424.

Mazzucato, M., & Penna, C. C. R. (2015). *Beyond market failures: The market creating and shaping roles of state investment banks* (Working paper). Levy Economic Institute of Bard College.

Mishra, D. K. (2013). Green strategies: Response of Indian banks to climate change. *The Ecoscan, 3,* 345–348.

Moon, J. (2007). The contribution of corporate social responsibility to sustainable development. *Sustainable Development, 15*(5), 296–306.

NGFS. (2019). *First comprehensive report: A call for action*. Climate Change as a Source of Financial Risk. Network for Greening the Financial System.

ONZ. (2015). *Transforming our world: The 2030 agenda for sustainable development*. United Nations 2015, A/RES/70/1. https://sdgs.un.org/sites/default/files/publications/21252030%20Agenda%20for%20Sustainable%20Development%20web.pdf.

Osborn D., Amy Cutter A., & Ullah, F. (2015). *Universal sustainable development goals: Understanding the transformational challenge for developed countries, Stakeholder Forum*. https://sustainabledevelopment.un.org/content/documents/1684SF_-_SDG_Universality_Report_-_May_2015.pdf.

Park, H., & Kim, J. D. (2020). Transition towards green banking: Role of financial regulators and financial institutions. *Asian Journal of Sustainability and Social Responsibility, 5*(5), 1–25. https://doi.org/10.1186/s41180-020-00034-3.

Pearce, D. W., & Atkinson, G. D. (1993). Capital theory and the measurement of sustainable development: An indicator of "weak" sustainability. *Ecological Economics, 8*(2), 103–108.

Pellicano, M., Vesci, M., Troisi, O., Cosimato, S., & Tuccillo, C. (2014). *Stakeholder engagement for sustainable development in banking industry: An Italian case study*. In 13th EBES conference—Istanbul, Turkey, pp 1–17. Available at http://www.researchgate.net/publication/275887421_Stakeholder_engagement_for_sustainable_development_in_banking_industry._An_Italian_case_study.

Rai, R., Kharel, S., Devkota, N., & Paudel, U. R. (2019). Customers perception on green banking practices: A desk. *Journal of Economic Issues, 10*, 82–95.

Russo, M. V., & Fouts, P. A. (1997). A resource-based perspective on corporate environmental performance and profitability. *The Academy of Management Journal, 40*(3), 534–559.

San-Jose, L., Retolaza, J. L., & Gutierrez, J. (2011). Are ethical banks different? A comparative analysis using the radical affinity index. *Journal of Business Ethics, 100*(1), 151–173.

Schoenmaker, D., & Tilburg, R. V. (2016). What role for financial supervisors in addressing environmental risks? *Comparative Economic Studies, 58*, 317–334.

Stern, N. (2016). *Climate change and central banks*. Bank for International Settlements.

Strategy for Smart, Sustainable and Inclusive Growth. European Commission, Communication of the European Commission 2020, Brussels, 3.3.2010, KOM(2010) 2020 final. www.ec.europa.eu/eu2020/pdf/1_PL_ACT_part1_v1.pdf.

Su, H., Rabbani, M. R., Ahmad, N., Sial, M. S., Cheng, G., Zia-Ud-Din, M., & Fu, Q. (2020). CSR, co-creation and green consumer loyalty: Are green banking initiatives important? A moderated mediation approach from an emerging economy. *Sustainability, 12,* 10688. https://doi.org/10.3390/su122410688.

Tara, K., Singn, S., & Kumar, R. (2015). Green banking for environmental management: A paradigm shift. *Current World Environment, 10*(3), 1029–1038.

UE. (2018). *Communication from the commission to the European Parliament, the European Council, the Council, the European Central Bank, the European Economic and Social Committee and the Committee of the Regions Action Plan: Financing sustainable growth, document 52018DC0097.* COM/2018/097 final. https://eur-lex.europa.eu/legal-content/PL/TXT/?uri=COM:2018:0097:FIN.

UNEP. (2016). *Definitions and concepts: Background note.* UNEP.

Volz, U. (2017). *On the role of Central banks in enhancing green finance, United Nations environment programme* (Inquiry Working Paper 17/01).

Weber, O. (2016, June). *The impact of green banking: Guidelines on the sustainability performance of banks* (pp. 1–7). The Chinese Case Policy Brief, 79.

Werner, R. A. (2003). *Princes of the Yen: Japan's central bankers and the transformation of the Economy.* Eastgate Books.

Windsor, D. (2006). Corporate social responsibility: Three key approaches. *Journal of Management Studies, 43*(1), 93–114.

Wood, D. J. (1991). Corporate social performance revisited. *Academy of Management Review, 16,* 691–718.

Wu, H.-C., & Cheng, C.-C. (2018). An empirical analysis of green experiential loyalty: A case study. *Journal of International Food and Agribusiness Marketing, 31,* 69–105.

Yip, A. W. H., Bocken, N. M. P. (2018). Sustainable business model archetypes for the banking industry. *Journal of Cleaner Production, 174,* 150–169.

Zaman G. (2012). Challenges and requirements for sustainable development of Romania'S agriculture based on the input-output analysis. *Romanian Journal of Economics, 35*(2(44)), 5–15.

Zimmermann, S. (2019). Same same but different: How and why Banks approach sustainability. *Sustainability, 11,* 2267. https://doi.org/10.3390/su11082267.

Ziolo, M., Filipiak, B. Z., Bąk, I., Cheba, K., Tîrca, D. M., & Novo-Corti, I. (2019a). Finance, sustainability and negative externalities. An overview of the European context. *Sustainability, 11*(15), 4249. https://doi.org/10.3390/su11154249.

Ziolo, M., Filipiak, B. Z., Bąk, I., & Cheba, K. (2019b). How to design more sustainable financial systems: The roles of environmental, social, and governance factors in the decision-making process. *Sustainability, 11*(20), 5604. https://doi.org/10.3390/su11205604.

CHAPTER 6

Sustainable Business Models of Companies

6.1 Introduction

Sustainable business models systematically integrate sustainability into business activities (Bocken et al., 2014). They are a development of the solutions applied so far, considering the systemic, axiological, and ontological approaches (Morioka et al., 2017) while matching market and economic trends. This approach supports the implementation of the concept of sustainable development in organizations, also through the adaptation of individual principles within the processes performed. The literature points to the possibility of transforming the principles of challenges and the principles of sustainable development into the competitive advantages of an organization (Belz & Binder, 2017). Sustainability principles can contribute to the creation of added value, which should be unambiguously identified as the sustainable value of an organization.

6.2 Definitions, Areas, and Frameworks of Sustainable Business Models

By creating new business models or evaluating the previously implemented strategies, the principles of sustainable development allow for the realization of not only economic, but also social and environmental goals; the proactive involvement of various stakeholders in the realization of the organization's goals; and viewing the effects of the organization's

© The Author(s), under exclusive license to Springer Nature Switzerland AG 2021

M. Ziolo et al., *Sustainability in Bank and Corporate Business Models*, Palgrave Studies in Impact Finance,
https://doi.org/10.1007/978-3-030-72098-8_6

from a long-term perspective to take into account one of the main principles of sustainable development: care for future generations (Morioka et al., 2017). The activities of organizations and their strategies should be oriented toward positive effects in the long term, which do not harm, and even support, the environment. This is possible through a holistic and systemic approach. Creating sustainable business models and implementing the principles of sustainable development are more than just providing value for the customer and realizing economic value (Yang et al., 2017).

Sustainable business models (SBMs) can be defined in different ways and the principles of sustainable business models can help create value: economic, social, and environmental (Rauter et al., 2017). SBMs incorporate proactive multi-stakeholder management, the creation of monetary and nonmonetary value for a broad range of stakeholders, and a long-term perspective (Geissdoerfer et al., 2017). In general, therefore, sustainable business models include the creation of economic, social, and environmental values. The economic value includes operating profits, while also realizing the value in use (Bowman & Ambrosini, 2000). It simultaneously creates value for the customer, i.e., the benefits perceived by the customer, and for the organization through market transactions. Environmental value means the company's impact on the environment and capital (Bocken et al., 2014). Social value includes elements that are considered of value by society and individuals, often being associated with psychological value (den Ouden, 2012). This approach means that business models must lead to potential economic value creation and have the potential to create value in a broader sense, i.e., positive net benefits from environmental and social perspectives (Laukkanen & Tura, 2020).

Sustainability in such business models should be viewed not only from the economic and financial points of view, but also from a broader socio-environmental perspective. There is a need to explore the links between business models and aspects of sustainability that exceed customer value and economic value (Lüdeke-Freund et al., 2019). Environmental value means the organization's impact on the environment (Stubbs & Cocklin, 2008). Social aspects are elements considered valuable by both society or individuals. They may be related to prosperity or, for example, to happiness. They relate to the psychological aspects of values (den Ouden, 2011). Creating value in the area of sustainability also requires eliminating and preventing the potential negative effects of introducing new strategies, e.g., related to conflicts of values or the necessity to make compromises (Tura et al., 2019).

Business models need to adapt their sustainability strategies to match business objectives with generally defined sustainability principles and challenges. They are the starting point for creating new, innovative, and sustainable business models. The aim of such strategies is to pay more attention to resource efficiency, reuse, responsible use of all resources, and elimination of harmful effects on the environment as part of business processes. From a social perspective, these models have increased interest in and care for the health and safety of employees, customers, and communities in general. It is important to comply with the rules and conditions for the organization to function ethically; to respect employee rights; to apply the principles of corporate social responsibility, fair remuneration, and non-discrimination; to engage in fair business practices and such competition; and to observe ethical principles (United Nations, 2019). Generalizing the considerations, it is in these three areas (Table 6.1) that solutions should be sought to help implement the new business model.

Consensus is lacking in the literature on the use of a single term for a business using sustainable principles to manage its business. Referring to Marioka et al. (2017), the literature in the context of implementing the principles of sustainable development uses a variety of terms in business strategy, e.g., business model for sustainability (Schaltegger et al., 2016), sustainability business model (Stubbs & Cocklin, 2008), flourishing/strongly sustainable business model (Upward & Jones, 2016), truly sustainable business models (Dyllick & Muff, 2016), and normative business model (Randles & Laasch, 2016). Despite the semantic differences in the presented terms, all of them are linked by the need to change the perspective of creating and running a business using the principles of sustainable development in a global context. Marioka et al. (2017) proposed basing the sustainable business model on three components (Richardson, 2005), which have also traditionally been used to create classic organizations that did not have a sustainability aspect in their strategy: value proposition (product/service, customer segments, and relationships), value creation and delivery system (key activities, resources, technologies, etc.), and value capture (cost structure and revenue streams) (Table 6.2). This approach points to the connection between business purpose and employees' values and beliefs, proactive and clear engagement in solving sustainability problems, and the need for system-level

Table 6.1 Sustainable value creation—elements and the conceptual framework

Environmental	*Social*	*Economic*
Increasing resource efficiency Reuse of products, by-products, and materials. Elimination/reduction of waste. Use of renewables (e.g., energy, raw materials)	**Safeguarding health and safety** The health and safety of employees/customers/communities are ensured	**Increasing cost-efficiency** Increases in efficiency and reduced costs compared to alternatives
Responsible use of resources Responsible use of natural resources (e.g., water, raw materials), respecting the welfare of ecosystems, people, and animals. No creation of rebound effects	**Respecting law, regulations, and rights** Laws, standards, and regulations (e.g., taxes, terms of use) and individuals' rights (e.g., privacy) are respected	**Increasing profits and business opportunities** Increases in profits and/or creation of new business opportunities and markets
No harmful environmental impacts and emissions No emissions (e.g., greenhouse gases) harming people or the environment. No harm to ecosystems or the environment	**Respecting employee, stakeholder, and individual rights** Employees' and stakeholders' terms are handled fairly (e.g., via paying living wages and non-discrimination). Equal treatment of employees/stakeholders/individuals	**Operational stability and risk reduction** Increases in long-term stability and risk reduction
Increasing environmental well-being Increases in biodiversity and environmental well-being by repairing previous damage and solving environmental problems (e.g., reducing ozone depletion)	**Ethical principles and no harmful social impact** Operations, products, and services do not harm people or communities. Human rights are respected (e.g., no child labor). Ethical principles are followed (e.g., caring use of resources, honest competition)	**Increasing attractiveness** Increases in reputation and brand value (e.g., attractiveness as an employee/collaborative partner)
	Increasing social well-being Increases in socio-psychological welfare (e.g., happiness, social cohesion)	**Increasing economic well-being** Increases in socio-economic welfare (e.g., employment)

Source United Nations (2019)

Table 6.2 Areas and aspect of SBM—sustainable business models

Sustainable business models		
Aspects		*Synthesis*
Value proposition	Products and services Social-environmental benefits Market segments and customer relationship "Represents the expected economic, environmental and social value added by the organizations" Carayannis et al. (2015), "Value proposition includes products and services" Osterwalder and Pigneur (2010), "Addresses short, medium and long term aspects of sustainability" Bansal and Des Jardine (2006), Lozano (2008)	SBM's value proposition summarizes the organizations' meaning of existence in terms of the sustainable value it aims to create and deliver, includes offerings (products and services) and seeks to address short-, medium- and long-term aspects

(continued)

Table 6.2 (continued)

Sustainable business models		
Aspects		*Synthesis*
Value creation and delivery system	Distribution channel Value configuration (key activities) Core competency Strategic partner Firm's resources, capabilities, and inter-organizational network (Richardson, 2009) Sustainability initiatives can be incorporated into operations/production, management, strategy, organizational systems, marketing and procurement, assessment, and communication (Lozano, 2012). These initiatives include: sustainable supply chain management (Carter & Rogers, 2008; Seuring & Müller, 2008), ecodesign (Brones & Carvalho, 2015; Luttropp & Lagerstedt, 2006), sustainable operations management (Kleindorfer et al., 2005), sustainability reporting (Brown & Deegan, 1998; Hahn & Kühnen, 2013), sustainable work design and ergonomics (Bolis et al., 2014), among many others	SBM's value creation and delivery system represents how the organization manages its resources, capabilities, and partners to enable its sustainable value proposition

Sustainable business models

Aspects		*Synthesis*
Value capture	Costs Revenue Companies' value captures include cost structure and revenue model (Osterwalder et al., 2005) SBM's enable financial value captured by the organization, but also depletion and/or increases natural, social and economic capital beyond its boundaries (Schaltegger et al., 2016)	Each stakeholder captures sustainable value created and delivered by the focal organization
Competitive advantage of SBM	There are win-win solutions (Elkington, 1994), sweet spots (Savitz & Weber, 2007), and shared value creation (Porter & Kramer, 2011), by exploring corporate sustainability to build competitive advantage SBM can be a strategic choice to increase competitive advantage (Engert et al., 2016)	SBM have potential to promote competitive advantage and, at the same time, contribution to SDG's
Contribution to SDG's—Sustainable Development Goals	Institutional theory (DiMaggio & Powell, 1983) can serve to justify firms' engagement to sustainable development, given laws and regulations (coercive pressure), competitors imitation (mimetic pressure), or ethical motivations (normative pressures)	

(continued)

Table 6.2 (continued)

Sustainable business models		
Aspects		*Synthesis*
Context factors	Factors include: Natural environment and social general context (Kolk & Mauser, 2002), legislation (Griffith & Bhutto, 2008; Tan et al., 2010), industry-specific competitive dynamics and market (Grosvold et al., 2014; Siebenhüner and Arnold, 2007), public opinion (Siebenhüner & Arnold, 2007) and technology level (Fan et al., 2014)	The organization is a node of a complex system, so outside of its boundaries affects the inside

Source Morioka et al. (2017), Gao and Li (2020)

changes to enable successful SBMs (Zioło et al., 2020). When creating sustainable business models, it is necessary to seek synergies between corporate and sustainability goals. An entrepreneurial and proactive attitude contributes to the change in regulations to achieve increasingly better results in the implementation of sustainable development principles with simultaneous economic profits. Business model innovation for sustainability is characterized by uncertainty and ambiguity (Roome & Louche, 2016).

The framework to support the SBM implementation of organizations must be aligned with sustainability performance goals (Fig. 6.1). The first element refers to the value proposition, which describes what value is delivered to which clients as the basic strategy for attracting these

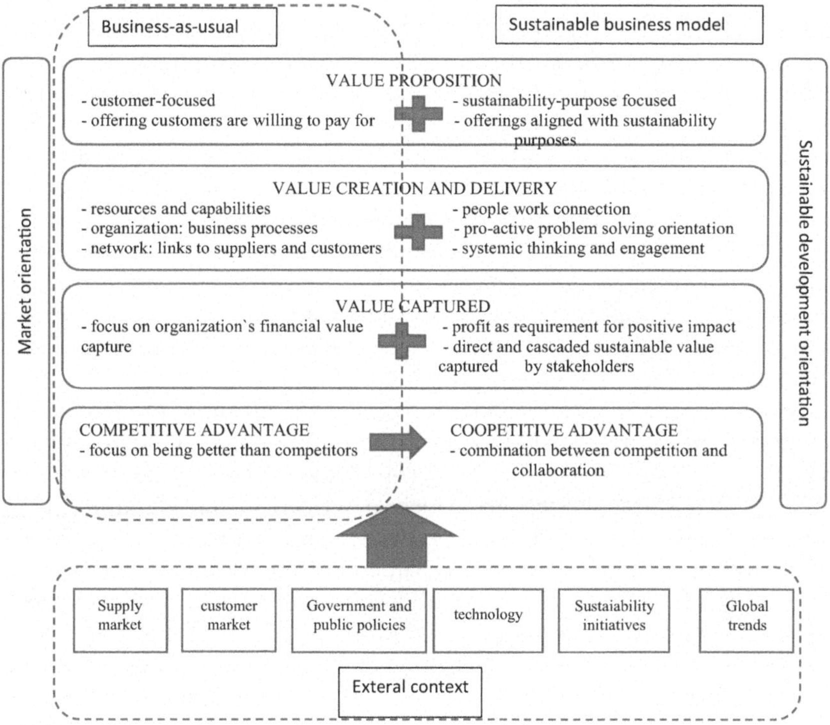

Fig. 6.1 From business to sustainable business model (*Source* [Morioka et al., 2017])

customers and creating a competitive advantage (Bocken et al., 2014). Value creation determines the sources of competitive advantage and rules that may enable the implementation of the value proposition (Aagaard, 2018). The value capture element explains how the firm will generate revenues and profits from its value proposition, accounting for its revenue stream, costs structure, and growth strategy (Bocken et al., 2014; Yang et al., 2017). Sustainable value principles (e.g., environmental, social, and economic value) are thus fully integrated into all three business model elements. This implies that sustainability is at the heart of the business model's value proposition; sustainable value is created and delivered to stakeholders and is transformed into economic value for the firm and its shareholders (Aagaard, 2018; Geissdoerfer et al., 2018; Schoneveld, 2020).

Research on sustainable business models is a relatively new area of interest (Baldassarre et al., 2017). The basis of sustainable models is appropriate adaptation to the specific business environment by implementing innovations to achieve the objectives of sustainable development (Gao & Li, 2020). The value proposition should be a promise to implement economic, social, and environmental aspects to the organization (Patala et al., 2016). The social–environmental benefits are an element of the value proposition in a sustainable business model, and thus the sustainable value proposition replaces the value proposition (Gao & Li, 2020). Value creation and delivery are key activities involving resources, competency, channels, and partners. Value capture is the revenue that a firm generates for itself from its value proposition, and the incurred costs (Abdelkafi & Täuscher, 2016) (Fig. 6.2).

Analyzing the literature and the sustainable business models of the framework, they are mostly based on the business model canvas (Osterwalder & Pigneur, 2010). The background and theoretical framework for sustainable business models are widely described in the literature. They contain different approaches and points of view, but focus on the implementation of the environmental, social, and economic aspects of business models (Bradley et al., 2020). In addition to the classic approach, they indicate solutions for applying innovations in such strategies and their connection to sustainability (Schaltegger et al., 2011). The following considerations are of major importance for the development of conceptualizing (Bocken et al., 2014; Lenssen et al., 2013) and developing tools for value mapping in sustainable business models, the aim of which is to develop value propositions, to indicate how stakeholders perceive

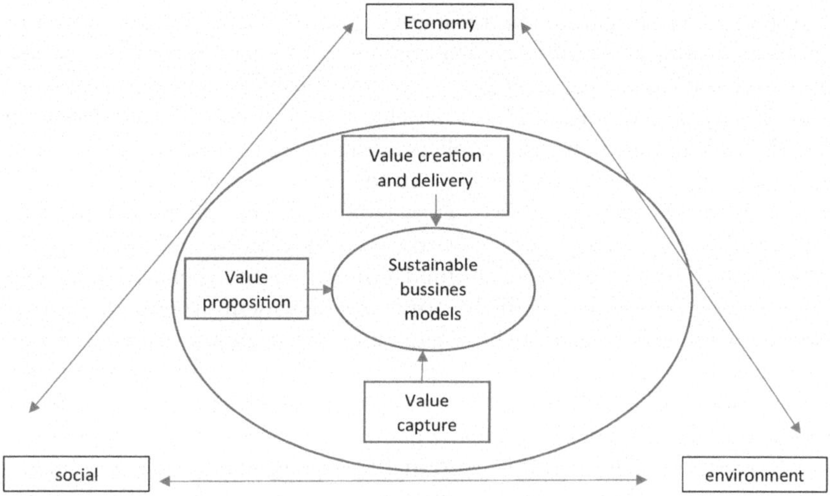

Fig. 6.2 Value in SBM (*Source* Own elaboration)

value in such models, highlighting the related created value, opportunities, and threats. These authors have also identified eight types of archetypes of sustainable business models. A different approach represents (Boons & Lüdeke-Freund, 2013) creating a framework to support structural research on sustainable development and sustainable enterprises in the context of focusing on innovations in such business models, presenting generic quality with vertical- and horizontal-structured relationships between different concepts. The research has been extended with the use of multi-method and multi-stage approaches, using expert research to identify and validate sustainable business models and create their taxonomy (Lüdeke-Freund et al., 2018). In the context of creating innovative business models, solutions based on the canvas business model are also built with complementary tools, methods, and concepts, based on case studies (França et al., 2017; Geissdoerfer et al., 2018). They offer a comprehensive overview of business models that consider elements of innovation, indicating the different steps of model creation. Morioka and de Carvalho (2016) examined the limitations and contributions of the performance measurement system to the theoretical framework of sustainable business models. The theoretical theses have also analyzed attempts to systematically integrate the principles of sustainable business development (including economic, environmental, and social objectives; multi-actor and long-term perspectives) into core business activities (Morioka et al., 2018). The theoretical basis has also aimed to expand knowledge on the relationship between sustainable product procurement and sustainable business models (Witjes & Lozano, 2016). Based on the business model canvas (Joyce & Paquin, 2016), knowledge has been extended to environmental and social aspects. Assessing the model of sustainable business, Upward and Jones (2016) examined companies in terms of designing the organization's activities in the context of balancing them. In another context, a model of relationship leadership for strategic sustainability in practice was developed, developing practices and opportunities to support sustainability in organizations in strategic terms (Kurucz et al., 2017). A tool was developed that helps organizations identify new opportunities for creating value through sustainable development. This is achieved using the life cycle analysis of captured and uncaptured value for key stakeholders (Yang et al., 2017). The model of Evans et al. (2017) references the understanding of innovation in the sustainable model, which contributes to better organization and implementation of economic, environmental, and social aspects.

As shown by an analysis of the literature, the theoretical framework has been created on the basis of the canvas business model. However, to a large extent, it has been extended. The most important elements of the extended sustainable business model approach include the following (Bradley et al., 2020):

- Extended framework:
    - Customer relationships and sensing,
    - Key stakeholders and partners,
    - Channels, value chain, and linkages;
    - Key resources,
- Builds on previous frameworks:
    - Value proposition and
    - Cost structure;
- Uses:
    - Customer segments and

  - Key activities;
- New category:
  - Value for individuals,
  - Value for organizations,
  - Value for society and the environment,
  - Use of a value proposition in context.

The model approach also allows the creation of various classifications of sustainable business models. It can help to creatively develop new models or to take inspiration from other existing solutions on the market to improve existing models by, for example, combining existing models (Lüdeke-Freund et al., 2018). New business models are not only about exploring relationships and elements related to the implementation of social, economic, and environmental aspects into organizational structures, but also about looking for alternative and more sustainable methods to design an architecture of new business solutions (Upward & Jones, 2016). The value of sustainable business models must not only relate to the principles of social, economic, and social aspects, but also consider short-term profits for the organization and long-term sustainability (Patala et al., 2016).

Various classifications and attempts to standardize sustainable business models can be identified in the literature. The types of models presented, among others, include Dohrmann et al. (2015), ideal-types (Stubbs & Cocklin, 2008), archetypes (Bocken et al., 2014), and pattern typologies (Lüdeke-Freund et al., 2018).

Dohrman et al. (2015) presented a division into four business models considering social aspects. Using the classic canvas model, they divided the models into Model I: one-sided social mission; Model II: two-sided social mission; Model III: market-oriented social mission; and Model IV: commercially used social mission. This breakdown and approach can be a first step toward creating sustainable business models. Using Model I, the central value proposition consists of the social mission, which is typically oriented toward the satisfaction of a consumption need. In Model II, the second generic business model features a higher degree of monetization of social value creation. Social ventures in this category address two different social target groups: one on the consumption and one on the production side. The third generic business model features a social target group only on the production side, which, although a recipient of social value creation, does not provide its support free of charge, but becomes a paid resource in the production of new value propositions. On the consumption side, there is no social target group. Instead, the venture focuses on the consumption needs of a market target group. Hence, the required funds are increasingly replaced by market revenues, thus increasing the monetization of social value creation. The fourth generic class features ventures with the highest degree of monetization of social value creation; this category is mainly characterized by a commercially used social mission. In particular, a social target group on the consumption side is attracted by the social mission, and, as a select group, is then used as a resource input to satisfy specific consumption needs of a different market target groups. As a consequence, market revenues are generated and largely reduce or replace funds for the social mission.

From a different perspective, a classification was presented by Stubbs and Cocklin (2008). Creating a sustainable business model involves implementing various activities. The authors of this new business model stated that the most important activities to be undertaken include redefining the purpose of business; reporting financial, environmental, and social outcomes; the stakeholder view of the firm; the role of leadership, nature, and environmental sustainability; modifying the taxation system; and retaining and reinvesting local capital—an SBM that encompasses a system perspective. The characteristics of the model are presented in Table 6.3.

The theory related to business model patterns indicates that particular elements and patterns are parts of business models (Amshoff et al., 2015; Abdelkafi et al., 2013). It is therefore worth using them to create a new taxonomy of sustainable business models and to better understand innovative business solutions. The existence of a catalog of patterns allows for the better classification and ordering of knowledge in the examined area. An interesting approach in this respect was presented by Lüdeke-Freund et al. (2018). They described a holistic approach using very detailed research and a taxonomy of sustainable business models based on it, considering pattern theory and business model patterns in the context of their implementation into sustainable business models. This approach allows both the whole models and their individual parts to be evaluated against different criteria. Prototype models are created that can support organizations to indicate to what extent they can create value. In this approach (Lüdeke-Freund et al., 2018), a sustainable business model pattern describes an ecological, social, and/or economic problem that

Table 6.3 Characteristic of a sustainable business model

	Characteristic			
	Economic	*Environmental*	*Social*	*Multidimensional or holistic*
Structural attributes	External bodies that track performance of companies use a triple bottom line (TBL) approach	Threefold strategy: offsets (do no harm but make amends if you do), sustainable (do no harm), restorative (leave the world better than you found it)	Stakeholder engagement skills: understanding stakeholders' needs and expectations (being relevant to stakeholders)	Systems approach: cooperative business strategy and planning. collaborative model including supply chain, competitors, government agencies, communities
	Lobby industry and government for changes to taxation system and legislation to support sustainability	Closed-loop systems: responsible for product throughout its lifecycle	Educate stakeholders; "relentless" communication	TBL approach to measure organizational performance
	Keep capital local: local shareholders and investment in local sustainability initiatives	Implement a services model	Implement stakeholder consultation program	Institutionalise sustainability in the business: "relentless" communication, stakeholder education, leadership, champions, and align internal performance measures

(continued)

Table 6.3 (continued)

	Characteristic			
	Economic	*Environmental*	*Social*	*Multidimensional or holistic*
		Industrial ecosystems and stakeholder networks	Get "buy-in" from internal and external stakeholders	Demand-driven model, not supply-driven model (driven by what people need, not driven by companies trying to get people to buy more)
Cultural attributes	Profit is a means not an ends	Treat nature as a stakeholder	Stakeholder approach (managing the organization for the benefit of all stakeholders and not prioritizing shareholders' expectations above other stakeholders)	Medium- to long-term focus
	Business makes a profit to do something more		Alignment of stakeholder expectations	Reduction in consumption

	Characteristic			
	Economic	*Environmental*	*Social*	*Multidimensional or holistic*
	"Higher purpose" to business than making money		Sharing of resources (people, profits, and time) among stakeholders to achieve sustainable outcomes	
	Shareholders invest for social and environmental impact reasons as well as for financial reasons		Relationship building (trust, two-way loyalty, honesty, integrity, and fairness, equity)	
	Shareholders temper expectations for short-term financial returns			

Source Stubbs and Cocklin (2008)

arises when an organization aims to create value, and it describes the core of a solution to this problem that can be repeatedly applied in a multitude of ways, situations, contexts, and domains. A sustainable business model pattern also describes the design principles, value-creating activities, and their arrangements that are required to provide a useful problem–solution combination. The result of such an approach is to match the identified benchmarks with the classic approach to sustainability aspects, reflected as the sustainability triangle. Eleven benchmarks were identified that reflect and are implemented in the sustainability triangle, as shown in Fig. 6.2.

The taxonomy is described in Table 6.4, which contains not only groups but also subgroups and indicates the impact on the individual elements of the sustainability triangle (Fig. 6.3).

The individual groups of benchmarks are as follows (Lüdeke-Freund et al., 2018):

- Pricing and revenue patterns: patterns that primarily address the revenue model of a business model, i.e., how offerings are priced and revenues generated;
- Financing patterns: patterns that address the financing model within a business model, i.e., how equity, debt, and operating capital are acquired;
- Ecodesign patterns: patterns that integrate ecological aspects into key activities and value propositions, i.e., how processes and offerings are designed to improve their ecological performance over their entire life cycle;
- Closing-the-loop patterns: patterns that help integrate the idea of circular material and energy flows into partnerships, key activities, and customer channels, i.e., how materials and energy flow into, out of, and return to a company;
- Supply chain patterns: patterns that modify the upstream (partners, resources, and capabilities) and/or downstream (customers, relationships, and channels) components of a business model, i.e., how inputs are sourced and target groups are reached;
- Giving patterns: patterns that help donate products or services to target groups in need, i.e., how costs are covered and social target groups are reached;
- Access provision patterns: patterns that create markets for otherwise neglected target groups, involving modified value propositions,

Table 6.4 The sustainable business model pattern taxonomy

Primary associated value creation (group)	*SBM pattern groups*	*Associated value creation (group and pattern)*
Mainly economic	Pricing and Revenue Patterns	Primary: mainly economic/Secondary: social–economic
	"Differential pricing"	Social–economic
	"Freemium"	Social–economic
	"Innovative product financing"	Mainly economic
	"Subscription model"	Mainly economic
	9. Service & Performance Patterns	Primary: mainly economic/Secondary: ecologic–economic
	"Pay for success"	Mainly economic
	"Product-oriented services"	Mainly economic
	"Result-oriented services"	Ecologic–economic
	"Use-oriented services"	Ecologic–economic
Social–economic	7. Access Provision Patterns	Primary: social–economic/Secondary: multiple
	"Building a marketplace"	Social–economic
	"E-transaction platforms"	Social–economic
	"Experience-based customer credit"	Social–economic
	"Last-mile grid utilities"	Social–economic
	"Value-for-money degrees"	Social
	"Value-for-money housing"	Mainly social
	10. Cooperative Patterns	Primary: social–economic/Secondary: none
	"Cooperative ownership"	Mainly economic
	2. Financing patterns	Primary: social–economic/Secondary: mainly economic
	"Crowdfunding"	Mainly economic
	"Microfinance"	Social–economic
	"Social business model: no dividends"	Social–economic
Social	6. Giving Patterns	Primary: social/Secondary: none

(continued)

Table 6.4 (continued)

Primary associated value creation (group)	*SBM pattern groups*	*Associated value creation (group and pattern)*
	"Buy one, give one"	Social
	"Commercially utilized social mission"	Social
	8. Social Mission Patterns	Primary: social/Secondary: social–economic
	"Expertise broker"	Social
	"Market-oriented social mission"	Social
	"One-sided social mission"	Social
	"Social business model: empowerment"	Social–economic
	"Two-Sided Social Mission"	Social
Mainly ecological	3. Ecodesign Patterns	Primary: mainly ecological/Secondary: ecologic–economic
	"Hybrid model/Gap-exploiter model"	Ecologic–economic
	"Maximise material productivity and energy efficiency"	Mainly ecological
	"Product design"	Mainly ecological
	"Substitute with renewables and natural processes"	Mainly ecological
	4. Closing-the-Loop Patterns	Primary: mainly ecological/Secondary: ecologic–economic
	"Co-product generation"	Ecologic–economic
	"Industrial symbiosis"	Ecologic–economic
	"Online waste exchange platform"	Ecologic–economic
	"Product recycling"	Ecologic–economic
	"Remanufacturing/Next life sales"	Mainly ecological
	"Repair"	Mainly ecological
	"Reuse"	Mainly ecological
	"Take back management"	Mainly ecological
	"Upgrading"	Mainly ecological

(continued)

Table 6.4 (continued)

Primary associated value creation (group)	*SBM pattern groups*	*Associated value creation (group and pattern)*
Integrative	11. Community Platform Patterns	Primary: integrative/Secondary: none
	"Sharing business"	Integrative
	5. Supply Chain Patterns	Primary: integrative/Secondary: multiple
	"Green supply chain management"	Mainly ecological
	"Inclusive sourcing"	Mainly social
	"Micro distribution and retail"	Social–economic
	"Physical to virtual"	Mainly economic
	"Produce on demand"	Mainly economic
	"Shorter supply chains"	Integrative

Source Lüdeke-Freund et al. (2018)

channels, revenue, pricing, and cost models, i.e., how value propositions are designed and delivered, and to whom;

- Social mission patterns: patterns that integrate social target groups in need, including otherwise neglected groups, either as customers or productive partners, i.e., how customers, partners, and employees are defined and integrated;
- Service and performance patterns: patterns that emphasize the functional and service value of products and that offer performance management, i.e., how value propositions are defined and delivered;
- Cooperative patterns: patterns that integrate a broad range of stakeholders as co-owners and co-managers, how partners are defined, and how the organization is governed;
- Community platform patterns: patterns that substitute resource or product ownership with community-based access to resources and products, and how value propositions are defined and delivered.

Each of the groups consists of successive subgroups, which describe in detail the indicated dependencies and elements. The taxonomy is described in Table 6.4, which contains not only groups but also

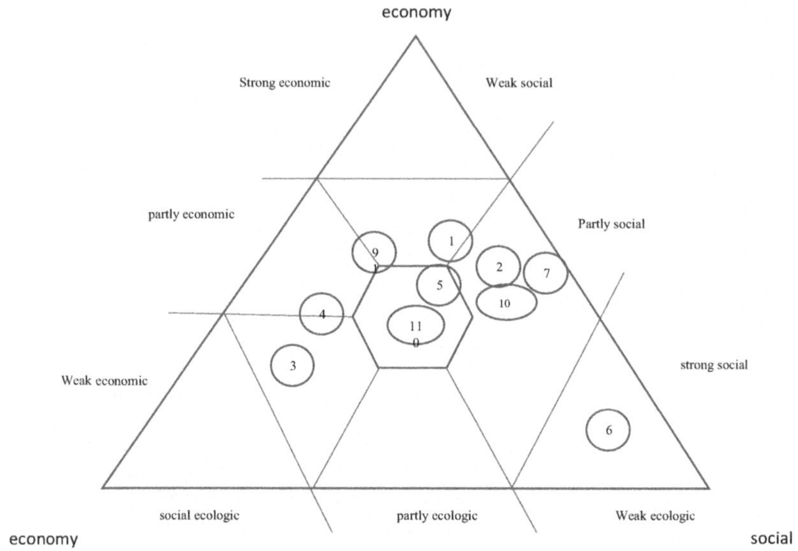

Pattern group	Forms of value creations
1-Pricing & revenue	a-economic
2-financing	b-social
3-ecodesign	c-ecological
4-closing-the-loop	d-mainly economic
5-supply chain	e-social economic
6-giving	f-mainly social
7-access provision	g-social-ecological
8-social mission	h-mainly ecological
9-service&performance	i-ecologic-economic
10-cooperative	j-integrative
11-community platform	

Fig. 6.3 A sustainable business model pattern (*Source* Lüdeke-Freund et al. [2018])

subgroups, and indicates the impact on the individual elements of the sustainability triangle.

This taxonomic approach allows for a better understanding of which elements create added value, to what extent, and in what area. It provides an opportunity to deeply understand the implementation of each element and create an innovative, sustainable business model.

When considering different classifications and model approaches to sustainable business strategies, circular strategies combined with sustainable strategies considered beneficial in equating positive impacts in the three dimensions of sustainability should also be taken into account (Geissdoerfer et al., 2017).

Sustainable business models cover a wide range of expected results and effects of implementation, the most important of which are better stakeholder relations, higher profitability, and better and more efficient use of resources, such as greater resistance to internal and external interference.

6.3 Drivers and Archetypes of the Sustainable Business Models

Alongside other drivers (compliance, economic performance, social and environmental responsibility, innovation, risk and quality management, and reputation), competitive advantage has been shown to be an important motivation for further integration of corporate sustainability into business (Engert et al., 2016).

Circular strategies have been categorized as elements of the archetypes of a sustainable business model. These archetypes have been combined with sustainable contributions and practices according to their nature and application, and how companies propose, create, deliver, and capture value (Barbieri & Santos, 2020):

1. Maximize material and energy efficiency (do more with less resources, generating less waste, emissions, and pollution);
2. Create value from waste (turn waste streams, emissions, and discarded products into feedstocks for other products and processes, and make best use of under-used capacity);
3. Deliver functionality, rather than ownership (provide services that satisfy users' needs without having to own physical products);
4. Encourage sufficiency (solutions that actively seek to reduce consumption and production);
5. Adopt a stewardship role (proactively engaging with all stakeholders to ensure their long-term health and well-being);
6. Repurpose the business for society and/or the environment (focusing the business on delivering social and environmental benefits, rather than economic profit maximization);

7. Integrate business in the community (integrating business back into local communities through employee ownership and collaborative approaches to business);
8. Develop scale-up solutions (delivering sustainable solutions on a large scale to maximize benefits for society and the environment);
9. Radical innovation (introduce system change through introduction of radical new technologies to facilitate a greener economy) (Short et al., 2012).

Drivers are an important element of implementing new, innovative business strategies. In general, they can be divided into external and internal. The motivating factors for implementing sustainable development are personal and value-based. In terms of creating motivators, both personal values and economic exigencies can often harmonize well with factors such as employee satisfaction and the availability of an interesting work environment, which can be viewed as internal drivers (Rauter et al., 2017). Internal drivers are related to operating cost benefits, revenue growth, brand integrity, effective internal champions, and organizational structures, which help embed sustainability into business processes (Kiron et al., 2012). A major role is played by leaders, especially in the initial phase of implementing sustainable business models. Their role is to define the model, provide a framework, and initiate sustainable development activities. Cooperation is necessary to implement the various elements of the strategy. Another important element is to accordingly plan the timing of activities. The internal driving forces include the culture of the organization. In this respect, the relationships that occur between all employees are extremely important, as are the style of leadership and involvement of employees, especially in terms of creating new ideas, providing support for new initiatives, and participating in decision-making. In the context of implementing and adapting the organizational culture, elements supporting the objectives of sustainable development can include health and safety audits, opinions of customers and suppliers, the degree of trust in the organization, flat and transparent organizational structure, employee development opportunities, constructive criticism, employees' own responsibility, commitment, continuous improvement, openness to innovation or risk-taking, and reduced staff turnover. Business models are the link between corporate strategies and operational activities (Rauter et al., 2017). This approach indicates that strategies are the driving force of the implementation of sustainable development.

The changing strategy consequently leads to changes in the business model and implementation of individual processes. Introducing a new model must also be connected with increasing public awareness of the principles of sustainable development and its role, promoting this idea, inscribing the principles of sustainable development into the strategic documentation, and basing the corporate policy on the above principles to help guarantee its success. The legal aspects should not be forgotten either. Leadership, organizational culture, and coherence between corporate strategy and the business model for sustainability have been identified as being relevant drivers in developing business models for sustainability. The need to maintain competition and competitive strength are not primary drivers in creating business models for sustainability. Although improving competitiveness and customer satisfaction are clearly relevant for companies, the desire for corporate sustainability is a primary driving force in this respect, consequently leading, again, to economic benefits (Rauter et al., 2017).

6.4 Tools to Assess the Functioning and Implementation of Sustainable Business Models

Reconfiguring a business strategy is difficult, which is why several tools have been created to support this process. Not all of them have been put into practice, but their knowledge facilitates designing innovative, sustainable business models. Such tools and processes include the design of flourishing enterprises approach (Upward & Jones, 2016), the triple-layered business model canvas (Joyce & Paquin, 2016), the value mapping tool (Evans et al., 2014; Lenssen et al., 2013), the value ideation concept (Geissdoerfer et al., 2016), and the Cambridge business model innovation process (Geissdoerfer et al., 2017). The tools are used to conceptualize a broad approach to creating new business models.

The design of flourishing enterprises approach (Upward & Jones, 2016) requests that sustainable business models include sustainability requirements. The presented ontological model, with clearly defined artifacts, was established from the literature using a comprehensive traditional literature review to establish the parameters for a subsequent systematic review (Jesson et al., 2011), indicating a framework of strongly sustainable business model (SSBM) propositions and principles. The framework consists of four formative propositions (FP1–4) and five instrumental principles (IP1–5). Formative propositions, which are part of this ontological model, contain the definitions of a strongly sustainable firm, value, a business model, and tri-profit. The instrumental principles include the concept, boundaries, and validation of an SSBM; the necessary financial viability of a business model; and modeling social benefits and environmental regeneration. The model is visualized in Fig. 6.4.

The model shows the relationships and connections between the most important elements that can create a business model. On this basis, the elements needed to implement sustainable concepts and create innovative business models can be identified.

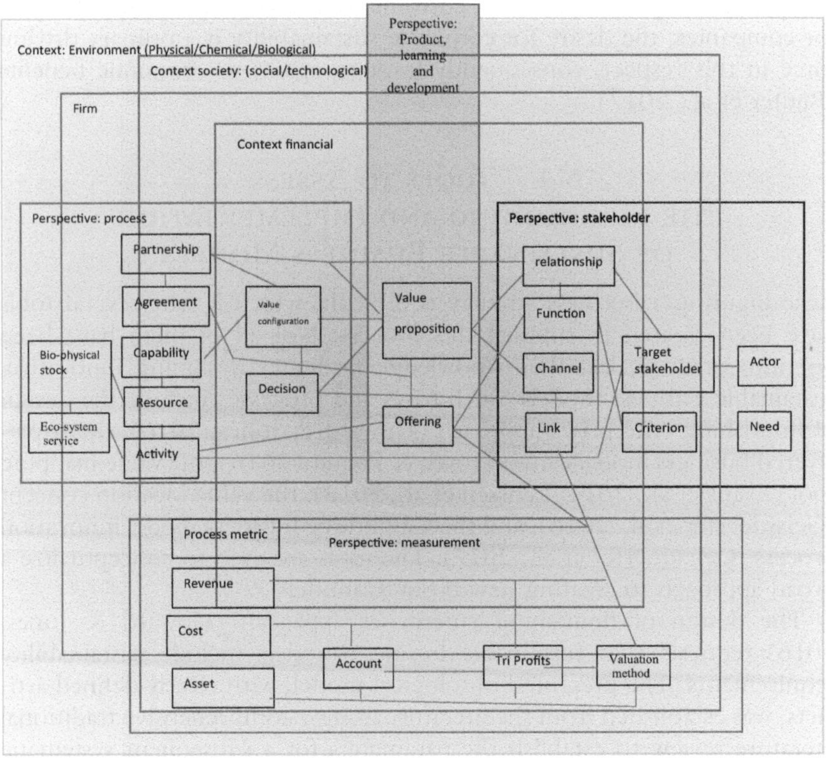

Fig. 6.4 Strongly sustainable business model ontology (*Source* Upward and Jones [2016])

Tools have also been created for mapping values in the range of different business models (Lenssen et al., 2013). Re-designing a business strategy can be the key to success and performance can be improved by implementing the concept of sustainable development. To create a sustainable business, a holistic view of the value proposition is required, which includes the benefits and costs to other stakeholders (besides customers and the firm) and specifically to society and the environment. A value mapping tool is conceived to help companies create value propositions to support sustainable business modeling. The tool adopts a qualitative approach to value analysis (Fig. 6.5). This tool has the following specific aims:

- Understand the positive and negative aspects of the value proposition of the value network (i.e., the network of stakeholders involved

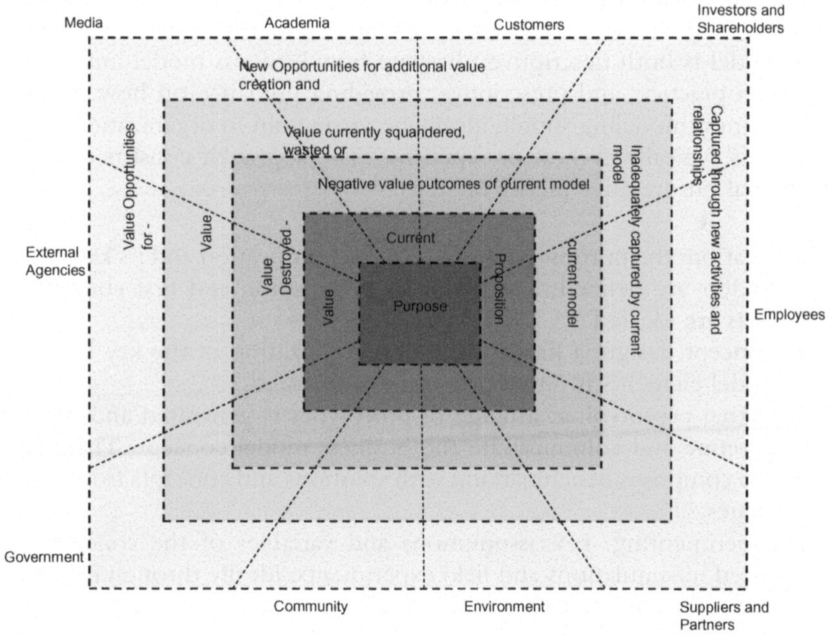

Fig. 6.5 The value mapping tool (*Source* On the basis of: Lenssen et al. [2013])

in the creation, delivery, and receipt of value associated with the provision of a product/service);
- Identify conflicting values (i.e., where one stakeholder benefit creates a negative for another stakeholder), so that action can be taken to tackle these;
- Identify opportunities for business model redesign and realignment of interests to reduce negative outcomes and improve the overall outcome for the stakeholders in the value network, especially for society and the environment.

The value mapping tool assists companies in embedding sustainability into the core of the business model through an improved understanding of the value proposition. It supports an iterative process for analyzing sustainable value creation opportunities from a multi-stakeholder perspective.

The Cambridge business model innovation process (Geissdoerfer et al., 2017) framework identifies opportunities for improvement, including by incorporating the social and environmental elements into the company. This model is both descriptive, showing how business model innovation occurs in practice, and prescriptive, providing guidance on how sustainable business modeling should ideally be carried out in organizations. The process is typically cyclical or repetitive. The approach consists of eight sequential but iterative phases or steps:

- Ideation: the purpose of business model innovation and its key stakeholders are defined, and the value proposition and first conceptual ideas are ideated.
- Concept design: a first rough conceptualization of the key business model elements is developed and documented.
- Virtual prototyping: a range of prototypes is generated and revised to refine and communicate the business model concept. The phase also comprises benchmarking with solutions and concepts from other parties.
- Experimenting: key assumptions and variables of the concept are tested in simulations and field experiments, ideally through randomized controlled trials.
- Detail design: an in-depth analysis and detailing of all the elements of the business model and interactions between these elements is conducted.

- Piloting: the entire concept is tested by first running a limited version of the business model in a subsection of the target market.
- Launch: the business model is rolled out across all responsible organizational units and the target market.
- Adjustment and diversification: the business model is revised according to initial plans, expectations, and strategic fit. Based on this evaluation, adjustments and diversifications are applied and, depending on the comprehensiveness of the necessary changes, the entire business model innovation process may be repeated.

The approach and its component elements are depicted in Fig. 6.6.

An example of a sustainable business model is a product service system (PSS), which shifts the business focus from designing, producing, and selling physical products to selling a system of products and services to fulfill specific client demands (Table 6.5).

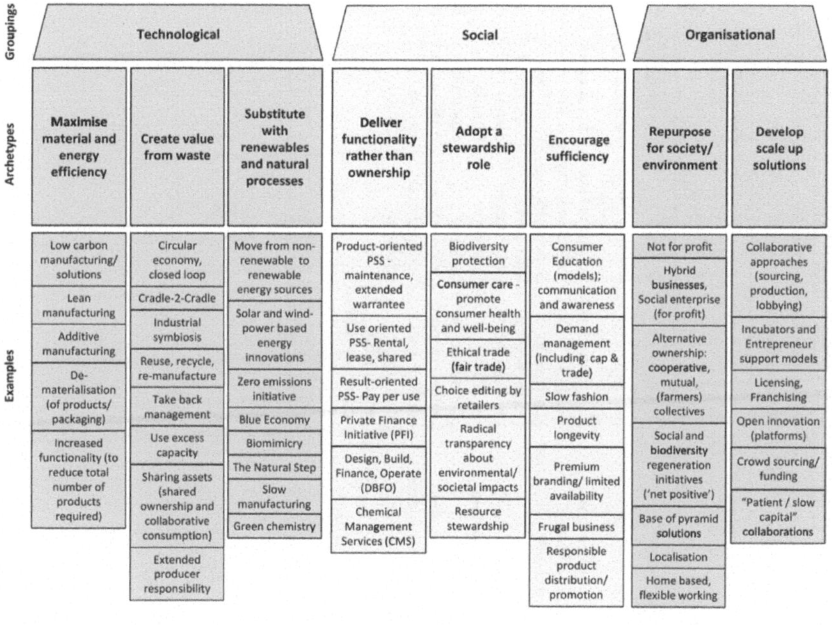

Fig. 6.6 Archetypes types of SBM (*Source* Boons and Lüdeke-Freund [2013])

Table 6.5 The Cambridge Business Model Innovation Process

The Cambridge Business Model Innovation Process

Phases	*Concept design*			*Detail design*			*Implementation*	
	1	*2*	*3*	*4*	*5*	*6*	*7*	*8*
Process	Ideation	Concept design	Virtual prototyping	Experimenting	Detail design	Piloting	Launch	Adjustment & Diversification
Activities	Vision/purpose formulation stakeholder definition value mapping/ideation sustainable value analysis Evaluation and selection of ideas	Integration of ideas Discussion of technological and general trends Definition of value creation, deliver and capture system/BM elements/BM dimensions	Benchmarking within industry Benchmarking with generic BM concept Prototype building Prototype building and selection	Identification od key variables Experiment design Running experiment Analysis and lessons learned	Detailed definition of all elements Overview of each elements Business transformation tools	Planning Implementation Analysis Adjustment Documentation and communication Identification of failure modes	Realization planning Implementation Scale up	Monitoring Reflection Adjustment Scale up Diversification Iteration of the business model innovation process

The Cambridge Business Model Innovation Process

Phases	*Concept design*			*Detail design*			*Implementation*	
	1	*2*	*3*	*4*	*5*	*6*	*7*	*8*
Challenges	Failed identification of opportunities Important stakeholder missed Failed the integrate top management from the beginning Lack of ambition/innovativeness	Insufficient mutual understanding Insufficient understanding of boundaries of the company capabilities to innovate Communications failures	Failed integration of important stakeholders into the process To much effort/prototypes become too big	No experiments Methodological issues Too much effort	Unsuited level of detail Missing of information from previous step Insufficient documentation Poor understanding of TMT risk	No pilots Unrealistic setting Too much effort	Insufficient information about failure modes Insufficient founding Inadequate timeframe/expectations Communications issues	Premature, too late or too little adjustment Unsuited diversification (agent motivation, missing core competencies, no ownership advantage)

Source Geissdoerfer et al. (2017)

6.5 A Catalog of Archetypes of Sustainable Business Models

Short et al. (2012) created a catalog of sustainable business model. When interpreting the data contained therein, the categories are based on (Boons & Lüdeke-Freund, 2013) the social, technological, and organizational aspects, and the possibilities of implementing innovations in this area. They are dominated by components related to the nature of the category.

Each of the archetypes can be considered in detail and its assumptions can be applied to the elements that create sustainable business models, i.e., value proposition, value creation, value delivery, and value capture. Organizations that are transforming their business models can use this directory of archetypes, and can choose one or more of the elements they consider most useful for the scope of their activities. The strategy innovation process should also include improving the archetype directory as well as a system for validating it. Archetypes have both potential and limitations to their use as they are based on historical data and examples of innovation (Boons & Lüdeke-Freund, 2013).

SBMs need to be continuously adjusted, improved, redesigned, and/or re-created to remain legitimate and relevant as political and economic contexts change, as societal and environmental needs and demands evolve, and as new opportunities for creating (further) sustainable value are identified (Girotra & Netessine, 2013; Evans et al., 2017). This points to the dynamic nature of SBMs (Schoneveld, 2020).

6.6 Conclusions

Creating sustainable business models is a difficult task, which requires the commitment of all stakeholders, the willingness for change expressed by owners and employees, as well as working with the different links in the chain to achieve the common good in sustainable business models. However, world trends, and a responsible look into the future necessitate the redesigning the organization's business and operations. This involves not only creating economic value, but also caring for the environment, responsible social involvement, and action for future generations, in accordance with the established principles of sustainable development. The indicated models, tools, and components should be helpful elements for managers.

References

Aagaard, A. (Ed.). (2018). *Sustainable business models: Innovation*. Implementation and Success: Springer.

Abdelkafi, N., & Täuscher, K. (2016). Business models for sustainability from a system dynamics perspective. *Organization & Environment, 29*(1), 74–96.

Abdelkafi, N., Makhotin, S., & Posselt, T. (2013). Business model innovations for electrical mobility—What can be learned from existing business model patterns? *International Journal of Innovation Management, 17*(01), 1340003

Amshoff, B., Dülme, C., Echterfeld, J., & Gausemeier, J. (2015). Business model patterns for disruptive technologies. *International Journal of Innovation Management, 19*(03), 1540002.

Baldassarre, B., Calabretta, G., Bocken, N. M. P., & Jaskiewicz, T. (2017). Bridging sustainable business model innovation and user-driven innovation: A process for sustainable value proposition design. *Journal of Cleaner Production, 147,* 175–186.

Bansal, P., & DesJardine, M. R. (2014). Business sustainability: It is about time. *Strategic Organization, 12*(1), 70–78.

Barbieri, R., & Santos, D. F. L. (2020). Sustainable business models and eco-innovation: A life cycle assessment. *Journal of Cleaner Production, 266.*

Belz, F. M., & Binder, J. K. (2017). Sustainable entrepreneurship: A convergent process model. *Business Strategy and the Environment, 26*(1), 1–17.

Bocken, N. M., Short, S. W., Rana, P., & Evans, S. (2014). A literature and practice review to develop sustainable business model archetypes. *Journal of cleaner production, 65,* 42–56.

Bolis, I., Morioka, S. N., & Sznelwar, L. I. (2014). When sustainable development risks losing its meaning: Delimiting the concept with a comprehensive literature review and a conceptual model. *Journal of Cleaner Production, 83,* 7–20.

Boons, F., & Lüdeke-Freund, F. (2013). Business models for sustainable innovation: State-of-the-art and steps towards a research agenda. *Journal of Cleaner Production, 45,* 9–19.

Bowman, C., & Ambrosini, V. (2000). Value creation versus value capture: Towards a coherent definition of value in strategy. *British journal of management, 11*(1), 1–15.

Bradley, P., Parry, G., & O'Regan, N. (2020). A framework to explore the functioning and sustainability of business models. *Sustainable Production and Consumption, 21,* 57–77.

Brones, F., & de Carvalho, M. M. (2015). From 50 to 1: Integrating literature toward a systemic ecodesign model. *Journal of Cleaner Production, 96,* 44–57.

Brown, N., & Deegan, C. (1998). The public disclosure of environmental performance information—A dual test of media agenda setting theory and legitimacy theory. *Accounting and Business Research, 29*(1), 21–41.

Carayannis, E. G., Sindakis, S., & Walter, C. (2015). Business model innovation as lever of organizational sustainability. *The Journal of Technology Transfer, 40*(1), 85–104.

Carter, C. R., & Rogers, D. S. (2008). A framework of sustainable supply chain management: Moving toward new theory. *International Journal of Physical Distribution & Logistics Management, 38*(5), 360–387. https://doi.org/10.1108/09600030810882816.

DiMaggio, P. J., & Powell, W. W. (1983). The iron cage revisited: Institutional isomorphism and collective rationality in organizational fields. *American Sociological Review*, 147–160.

den Ouden, E. (2011). *Innovation design: Creating value for people, organizations and society*. Springer Science & Business Media.

den Ouden, E. (2012). *Innovation design: Creating value for people*. Springer, Organizations and Society.

Dohrmann, S., Raith, M., & Siebold, N. (2015). Monetizing social value creation—A business model approach. *Entrepreneurship Research Journal, 5*(2), 127–154.

Dyllick, T., & Muff, K. (2016). Clarifying the meaning of sustainable business: Introducing a typology from business-as-usual to true business sustainability. *Organization & Environment, 29*(2), 156–174.

Elkington, J. (1994). Towards the sustainable corporation: Win-win-win business strategies for sustainable development. *California Management Review, 36*(2), 90–100.

Engert, S., Rauter, R., & Baumgartner, R. J. (2016). Exploring the integration of corporate sustainability into strategic management: Aliterature review. *Journal of Cleaner Production, 112*, 2833–2850.

Evans, S., Rana, P., & Short, S. W. (2014). Final set of tools & methods that enable analysis of future oriented, novel, sustainable, value adding business-models and valuenetworks. Deliverable D2. 6, Project 262931, SustainValue: Sustainable value creation in manufacturing networks.

Evans, S., Vladimirova, D., Holgado, M., Van Fossen, K., Yang, M., Silva, E. A., & Barlow, C. Y. (2017). Business model innovation for sustainability: Towards a unified perspective for creation of sustainable business models. *Business Strategy and the Environment*, *26*(5), 597–608.

França, C. L., Broman, G., Robert, K. H., Basile, G., & Trygg, L. (2017). An approach to business model innovation and design for strategic sustainable development. *Journal of Cleaner Production, 140*, 155–166.

Gao, P., & Li, J. (2020). Understanding sustainable business model: A framework and a case study of the bike-sharing industry. *Journal of Cleaner Production, 267*, 122229.

Geissdoerfer, M., Bocken, N. M., & Hultink, E. J. (2016). Design thinking to enhance the sustainable business modelling process–A workshop based on a value mapping process. *Journal of Cleaner Production, 135,* 1218–1232.

Geissdoerfer, M., Morioka, S. N., de Carvalho, M. M., & Evans, S. (2018). Business models and supply chains for the circular economy. *Journal of Cleaner Production, 190,* 712–721.

Geissdoerfer, M., Savaget, P., & Evans, S. (2017). The Cambridge business model innovation process. *Procedia Manufacturing, 8,* 262–269.

Girotra, K., & Netessine, S. (2013). OM forum—business model innovation for sustainability. *Manufacturing & Service Operations Management, 15*(4), 537–544.

Griffith, A., & Bhutto, K. (2008). Improving environmental performance through integrated management systems (IMS) in the UK. *Management of Environmental Quality: An International Journal, 19*(5), 565–578.

Grosvold, J., Hoejmose, S. U., & Roehrich, J. K. (2014). Squaring the circle: Management, measurement and performance of sustainability in supply chains. *Supply Chain Management: An International Journal, 19*(3), 292–305.

Hahn, R., & Kühnen, M. (2013). Determinants of sustainability reporting: a review of results, trends, theory, and opportunities in an expanding field of research. *Journal of Cleaner Production, 59,* 5–21.

Jesson, J., Matheson, L., & Lacey, F. M. (2011). *Doing your literature review: Traditional and systematic techniques*. Sage.

Joyce, A., & Paquin, R. L. (2016). The triple layered business model canvas: A tool to design more sustainable business models. *Journal of cleaner production, 135*, 1474–1486.

Kiron, D., Kruschwitz, N., Haanaes, K., & von Streng Velken, I. (2012). Sustainability nears a tipping point. *MIT Sloan Management Review, 53*(2), 69–74.

Kleindorfer, P. R., Singhal, K., & Van Wassenhove, L. N. (2005). Sustainable operations management. *Production and Operations Management, 14*(4), 482–492.

Kolk, A., & Mauser, A. (2002). The evolution of environmental management: from stage models to performance evaluation. *Business Strategy and the Environment, 11*(1), 14–31.

Kramer, M. R., & Porter, M. (2011). *Creating shared value* (Vol. 17). FSG.

Kurucz, E. C., Colbert, B. A., Luedeke-Freund, F., Upward, A., & Willard, B. (2017). Relational leadership for strategic sustainability: Practices and capabilities to advance the design and assessment of sustainable business models. *Journal of Cleaner Production, 140*, 189–204.

Laukkanen, M., & Tura, N. (2020). The potential of sharing economy business models for sustainable value creation. *Journal of Cleaner Production, 253*, 120004.

Lenssen, G., Painter, M., Ionescu-Somers, A., Pickard, S., Bocken, N., Short, S., … & Evans, S. (2013). *A value mapping tool for sustainable business modelling*. Corporate Governance.

Lozano, R. (2008). Envisioning sustainability three-dimensionally. *Journal of Cleaner Production, 16*(17), 1838–1846.

Lozano, R. (2012). Orchestrating organisational changes for corporate sustainability. *Greener Management International, 57,* 43–64.

Lüdeke-Freund, F., Carroux, S., Joyce, A., Massa, L., & Breuer, H. (2018). The sustainable business model pattern taxonomy—45 patterns to support sustainability-oriented business model innovation. *Sustainable Production and Consumption, 15,* 145–162.

Lüdeke-Freund, F., Gold, S., & Bocken, N. M. (2019). A review and typology of circular economy business model patterns. *Journal of Industrial Ecology, 23*(1), 36–61.

Luttropp, C., & Lagerstedt, J. (2006). EcoDesign and the Ten Golden Rules: generic advice for merging environmental aspects into product development. *Journal of Cleaner Production, 14*(15–16), 1396–1408.

Morioka, S. N., & de Carvalho, M. M. (2016). A systematic literature review towards a conceptual framework for integrating sustainability performance into business. *Journal of Cleaner Production, 136,* 134–146.

Morioka, S. N., Bolis, I., & de Carvalho, M. M. (2018). From an ideal dream towards reality analysis: Proposing Sustainable Value Exchange Matrix (SVEM) from systematic literature review on sustainable business models and face validation. *Journal of Cleaner Production, 178,* 76–88

Morioka, S. N., Bolis, I., Evans, S., & Carvalho, M. M. (2017). Transforming sustainability challenges into competitive advantage: Multiple case studies kaleidoscope converging into sustainable business models. *Journal of Cleaner Production, 167*, 723–738.

Osterwalder, A. & Pigneur, Y. (2010). *Business Model Generation—A Handbook 131 for Visionaries, Game Changers and Challengers*. Wiley.

Osterwalder, A., Pigneur, Y., & Tucci, C. L. (2005). Clarifying business models: Origins, present, and future of the concept. *Communications of the Association for Information Systems, 16*(1), 1.

Patala, S., Jalkala, A., Keränen, J., Väisänen, S., Tuominen, V., & Soukka, R. (2016). Sustainable value propositions: Framework and implications for technology suppliers. *Industrial Marketing Management, 59,* 144–156.

Randles, S., & Laasch, O. (2016). Theorising the normative business model. *Organization & Environment, 29*(1), 53–73.

Rauter, R., Jonker, J., & Baumgartner, R. J. (2017). Going one's own way: Drivers in developing business models for sustainability. *Journal of Cleaner Production, 140,* 144–154.

Richardson, B. J. (2009). Keeping ethical investment ethical: Regulatory issues for investing for sustainability. *Journal of Business Ethics, 87*(4), 555–572.

Richardson, J. E. (2005). The business model: an integrative framework for strategy execution. *Available at SSRN 932998.*

Roome, N., & Louche, C. (2016). Journeying toward business models for sustainability: A conceptual model found inside the black box of organisational transformation. *Organization & Environment, 29*(1), 11–35.

Savitz, A. W., & Weber, K. (2007). The sustainability sweet spot. *Environmental Quality Management, 17*(2), 17–28.

Schaltegger, S., Hansen, E. G., & Lüdeke-Freund, F. (2016). *Business models for sustainability: Origins, present research, and future avenues.*

Schaltegger, S., Lüdeke-Freund, F., & Hansen, E. G. (2011). Business cases for sustainability and the role of business model innovation: Developing a conceptual framework. Centre for Sustainability Management (CSM), Leuphana Universität Lüneburg.

Schoneveld, G. C. (2020). Sustainable business models for inclusive growth: Towards a conceptual foundation of inclusive business. *Journal of Cleaner Production, 277*, 124062

Seuring, S., & Müller, M. (2008). From a literature review to a conceptual framework for sustainable supply chain management. *Journal of Cleaner Production, 16*(15), 1699–1710.

Short, S. W., Bocken, N. M. P., Rana, P., & Evans, S. (2012). Business model innovation for embedding sustainability: A practice-based approach introducing business model archetypes. In *10th Global Conference on Sustainable Manufacturing.*

Siebenhüner, B., & Arnold, M. (2007). Organizational learning to manage sustainable development. *Business Strategy and the Environment, 16*(5), 339–353.

Stubbs, W., & Cocklin, C. (2008). Conceptualizing a "sustainability business model". *Organization & Environment, 21*(2), 103–127.

Tan, A. R., Matzen, D., McAloone, T. C., & Evans, S. (2010). Strategies for designing and developing services for manufacturing firms. *CIRP Journal of Manufacturing Science and Technology, 3*(2), 90–97.

Tura, N., Keränen, J., & Patala, S. (2019). The darker side of sustainability: Tensions from sustainable business practices in business networks. *Industrial Marketing Management, 77*, 221–231.

United Nations. (2019). *About the sustainable development goals.* Available at https://www.un.org/sustainabledevelopment/sustainable-development-goals/. Accessed 12 2020.

Upward, A., & Jones, P. (2016). An ontology for strongly sustainable business models: Defining an enterprise framework compatible with natural and social science. *Organization & Environment, 29*(1), 97–123.

Witjes, S., & Lozano, R. (2016). Towards a more Circular Economy: Proposing a framework linking sustainable public procurement and sustainable business models. *Resources, Conservation and Recycling, 112,* 37–44.

Yang, M., Evans, S., Vladimirova, D., & Rana, P. (2017). Value uncaptured perspective for sustainable business model innovation. *Journal of Cleaner Production*, *140*, 1794–1804.

Yang, M., Vladimirova, D., & Evans, S. (2017). Creating and capturing value through sustainability: The sustainable value analysis tool a new tool helps companies discover opportunities to create and capture value through sustainability. *Research-Technology Management*, *60*(3), 30–39.

Zioło, M., Bąk, I., Cheba, K., & Spoz, A. (2020). Sustainable business models of enterprises-actual and declared activities for ensuring corporate sustainability. *Procedia Computer Science, 176,* 1497–1506.

CHAPTER 7

Business Models of Banks Toward Sustainability and ESG Risk

Environmental, social, and corporate governance (ESG) risk has particular relevance to banks. ESG risk can influence a bank's financial performance.[1] Several banks have incorporated ESG factors into the decision-making process and have adapted their business models toward sustainability. The aim of this chapter is to identify what ESG factors have been incorporated by banks in the risk assessment process? in this context to discuss sustainable business model archetypes in the banking sector, and identify challenges and prospects faced by banks in designing their business models toward sustainability.

7.1 Introduction

The ongoing demographic, climate, and socio-economic changes have their financial dimension and have created a new type of risk: environmental, social, and governance (ESG) risk, which is currently one of the most significant from the perspective of the assessment and functioning of the banking sector. It includes, in particular, risks such as environmental and reputational risk. The contemporary financial sphere is therefore influenced by a number of factors that determine the necessity

[1] The chapter is a part of the project and research results presented in M. Ziolo, Sustainable finance. Development, Risk, Market, PWE, Warsaw 2020 and Financial sphere toward sustainable development. Ed. K. Raczkowski, M. Ziolo, CeDeWu, Warsaw 2017.

© The Author(s), under exclusive license to Springer Nature Switzerland AG 2021

M. Ziolo et al., *Sustainability in Bank and Corporate Business Models*, Palgrave Studies in Impact Finance,
https://doi.org/10.1007/978-3-030-72098-8_7

of implementing changes and innovative solutions in the banking sector. Considering the economic, social, and environmental perspective that characterizes sustainable banking allows for the more adequate adaptation of banks to new challenges. The banking sector reacts by responding to new dilemmas and challenges with a certain delay in relation to the real sphere. However, noticeable changes in banking are occurring on many levels, related to the domestic, international, and global perspectives. The processes of change are occurring in all kinds of banking (retail, wholesale, green, and sustainable). In the banking sector, the sustainability concept is reflected in green and responsible banking. The offer of green banking services supports the process of adaptation and transformation toward a green economy, and responsible banking additionally implements the assumptions of the concept of social inclusion and respects and promotes ethical standards.

The concept of sustainable banking is disseminated through institutions and instruments used within the banking sector. Therefore, respecting the postulates of sustainable banking is also a challenge from the point of view of institutional economics. The organization of the banking sphere, including the banking system in terms of its adaptation to the requirements of sustainability, requires a number of changes of a structural and organizational nature, as well as in the organizational culture of individual institutions and their missions. These changes should be ordered and controlled, which is conditioned by the quality of regulations. The regulations are important primarily from the point of view of shaping and supervising the offer of financial products and services, but also the safety and stability of the banking sector, which depend on the quality of risk management (including the approach to ESG risk).

Expectations for the banking sector and its role are changing. The process of adjusting to the paradigm of sustainable banking is long-term and requires trade-offs. The key to its efficient implementation is a proper understanding of the need for changes by stakeholders. Sustainable banking, in turn, is an important element in the process of implementing the postulates of sustainable development and growth; in this context, they affect the indicators of achieving the goals in these spheres. The interdependence and adequacy of the real and financial spheres in shaping sustainable development and growth are therefore key and strategically important success factors for the implementation of the intentions of both concepts and for ensuring the quality functioning of states, economies, and societies from a global perspective.

The goal of this chapter is to present ESG factors incorporated by banks in the risk management process and to describe their impact on banks' business models. The remainder of the chapter is organized as follows: Sect. 7.2 discusses ESG risk issues, and concepts and factors of risk. In Sect. 7.3, sustainable banks' business models are described. Section 7.4 is the chapter's conclusions.

7.2 ESG Risk Management and Its Impact on Banking Sector

ESG risk is inherent in the category of sustainable finance and is one of the most significant research trends in this category. On the basis of the conducted research, the relationships between ESG risk, its determinants, and the rate of return on the investment portfolio are being increasingly emphasized, indicating neutral or positive correlations between the variables and the relationship between ESG risk determinants and the cost of capital. ESG risk is a phenomenon so complex and difficult to manage that its management requires more skills and competences from leadership than management of the other risk categories, thus determining better financial results and the effects of operations of enterprises active in sectors sensitive to ESG risk (Fullwiler, 2015, s. 5). The risk at the interface between the business and social environments, the natural environment, and the risk related to corporate governance is a category defined as ESG risk. Table 7.1 presents the ESG risk factors incorporated by financial institutions in the decision-making process for every group of risk: environmental, social, and governance.

The treatment of non-financial factors by groups of factors (environmental, social, and management) is heterogeneous and depends on the purpose of the classification. One of the classifications is the one that was prepared for the purposes of ISO 26000 (Table 7.2).

According to Deloitte global risk management survey, ESG risk is on the top list of risks of financial institutions in coming years (Fig. 7.1).

Considering the growing role of ESG risk in the environment of banking and financial institutions (especially in the context of environmental risk), one of the ways to minimize this type of risk is an efficiently operating system of its management. ESG risk directly affects the financial

Table 7.1 ESG risk factors from perspective of financial institutions

Risk	*Risk factor*
Environmental	Carbon intensity emissions, climate change, control of environmental impacts, ecodesign (financial green products and services), eco-efficiency, emissions, energy consumptions, environmental policy management, environmental reporting, environmental risk management, hazardous waste, materials recycled and reused, packaging, pollution management/recourses, protection of biodiversity, raw material sourcing, renewable energy consumption, travel and transport impact, waste management reduction, water use and management, industry-specific criteria
Social	Business behavior, community relations, corporate citizenship/philanthropy, customer relationship management, customer and product responsibility, diversity human capital development and training, human rights criteria, labor management, local suppliers, market ethics, non-discrimination, promotion equality, privacy and data security, protection of children, exclusion of children labor, quality of working condition, respect of trade unions, responsible investing, rights of indigenous people, social reporting, stakeholders engagement, supply chain management, talent attraction/retention, work–life balance, industry-specific criteria
Governance	Antitakeover policy, audit and control system, board diversity, board structure, brand management, business ethics and fraud, codes of conduct/compliance, corporate government functions and commitments, prevention of corruption and bribery, remuneration of members of the executive team, respect of shareholders rights, risk and crisis management, transparency, vision and strategy, antitrust policy, industry-specific criteria

Source E. Escrig-Olmedo, M. Ángeles Fernández-Izquierdo, I. Ferrero-Ferrero, J. M. Rivera-Lirio, & M. J. Muñoz-Torres. (2019). Rating the raters: Evaluating how ESG rating agencies integrate sustainability principles. *Sustainability, 11*, 915; M. Ziolo, B. Z. Filipiak, I. Bąk, & K. Cheba. (2019). How to design more sustainable financial systems: The roles of environmental, social, and governance factors in the decision-making process. *Sustainability, 11*, 5604

stability and liquidity of banks because the banks bear the financial consequences of the loss of financial liquidity of the entities to which they lend, and of the financial penalties imposed resulting from failure to comply with regulations and taking actions harmful to the natural environment. Another financial consequence for banks is the probability of losing the market value of legal collateral for the repayment of loans granted by them as a result of environmental pollution, among others. Contaminated land/real estate is difficult to sell at a market price, so will not be fully valuable for banks as creditors as a legal security for repayment of this type of real estate. Therefore, banking and financial institutions should

Table 7.2 ESG risk factors according to ISO 26000 typology

Risk	*Risk factor*
Environmental	Pollution prevention, sustainable resource use, mitigation and adaptation to climate change, environmental protection, biodiversity and restoration of natural habitats
Social	Due diligence, human rights situations, avoidance of complicity / assistance in illegal activities, complaint handling, discrimination and vulnerable groups, civil, economic, social and cultural rights, fundamental principles and rights at work, employment and industrial relations, working conditions and social protection, social dialogue, health and safety, human development and training in the workplace
Governance	Anti-corruption, responsible involvement in political activities, fair competition, promoting social responsibility in the value chain, respecting property rights

Source What Is ISO 26,000—Guidance on Social Responsibility? https://asq.org/quality-resources/iso-26000 (access 31.12.2020)

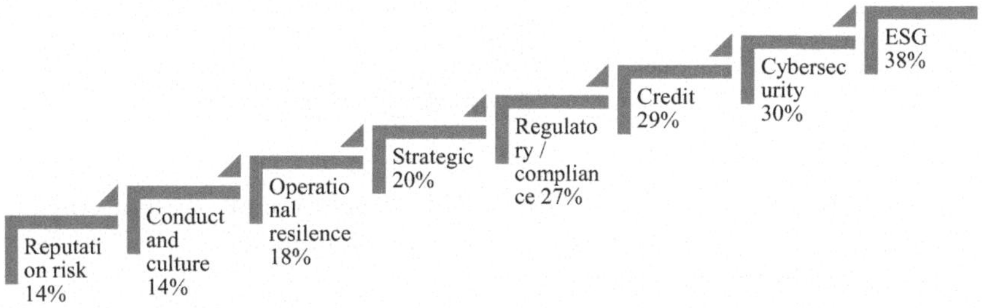

Fig. 7.1 Top list of financial institutions' risks (*Source* Own elaboration based on https://www.bloombergquint.com/onweb/esg-risks-top-the-list-of-near-term-concerns-for-bank-executives [access 31.12.2020])

redefine both the meaning and the approach to operational risk in their activities (Fig. 7.2).

- An effective ESG risk management system should include the following elements (Bank Track, 2006);
- preliminary diagnosis and assessment of the factors determining the key ESG risk in the operations in terms of both threats and opportunities for shaping the development policy of a banking and financial institution;

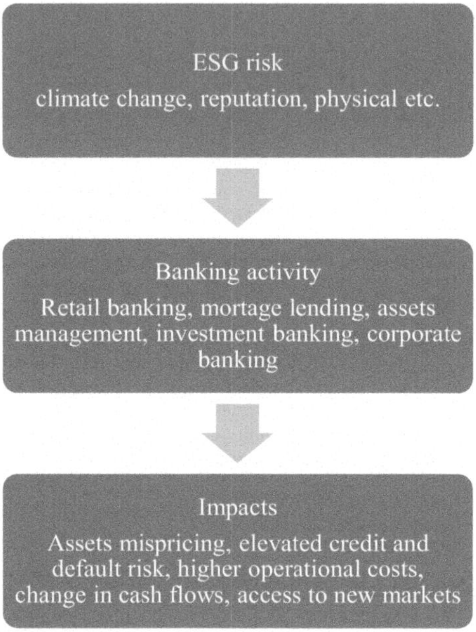

Fig. 7.2 ESG risk impact on banking activity (*Source* Own elaboration based on The Shareholder Association for Research and Education (SHARE), Banking on 2, 2015 and https://www.slideshare.net/saraninee/esg-risks-and-thai-banks-time-to-walk-the-talk [access 30.12.202])

- development of the bank's sustainable development concept considering ESG risk as well as regulations, recommendations, and best practices in its management;
- developing sectoral, regional, and national policies that are based on international standards and best practices, which contain clear standards and guidelines necessary to approve or reject financial transactions subject to ESG risk;
- shaping the organizational structure and assigning responsibility to individual units/employees for compliance of the activities carried out, their coherence, and the ESG risk management strategy; development of a sustainable reporting and reporting system supporting ESG risk management;
- assessment and diagnostics of own resources and staff awareness of the role and importance of the ESG risk management system in the bank's operations and in work positions held by individual persons, and the consequences of their decisions and actions in the context of the impact on the efficiency of this system;

- defining standards and procedures for financial transactions that include due diligence, consultation, and compliance;
- development and use of simple tools supporting the due diligence procedure that are understandable and usable by the bank's staff, clients, and stakeholders; these tools should allow for the assessment of the impact of ESG risk on financial products and services offered by the bank, the customer's track record, commitment to sustainable development, and the ability to achieve the expected environmental and social effects;
- providing an independent, comprehensive, and substantive review of all technical documents provided by the client;
- developing procedures and tools enabling the exchange of knowledge and information on the subject of potential clients with other entities, including other banks, governments, and rating agencies;
- introducing minimum standards and requirements in the field of ESG risk, which will allow for and provide decision support at the stage of deciding whether or not to cooperate with a given client by a financial institution (e.g., checklist, list of sectors of dirty sectors and dirty investments that are not in the sphere of interests and activities of a given financial institution);
- developing a system to monitor clients in terms of ESG risk and introducing minimum standards obliging clients to report on the ESG risk status in their operations;
- developing a mechanism for submitting complaints and grievances for stakeholders regarding the ESG risk management system;
- developing a policy of informing external entities about the ESG risk management system adopted by the banking or financial institution; and
- providing a monitoring and evaluation system for the ESG risk management.

ESG risk approach in banking sector presents in Fig. 7.3.

It is worth mentioning that banking hub distinguishes 7 ESG strategies (bankinghub, 2020):

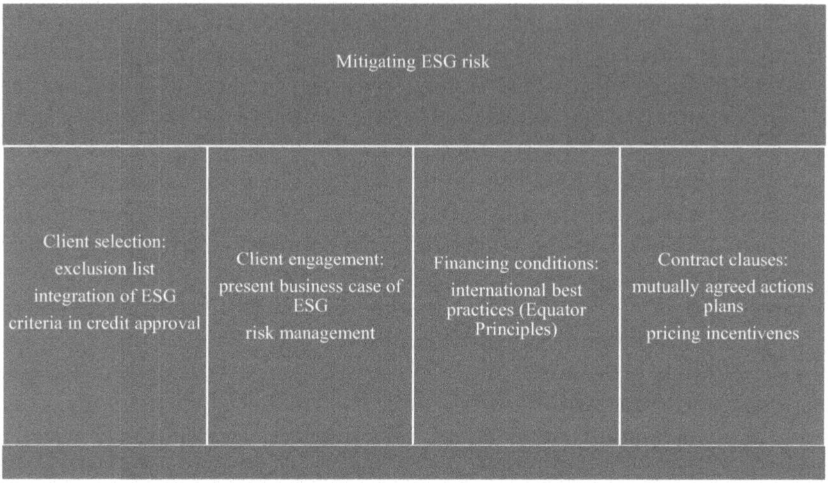

Fig. 7.3 Mitigating ESG risk in banking sector (*Source* Own elaboration based on https://www.slideshare.net/sarinee/esg-risks-and-thai-banks-time-to-walk-the-talk [access 18.02.2021])

- exclusions (screening of companies, sectors or countries involved in non ESG activities) and norm-based screening (exclusion ESG activities that don't meet ESG international standards) as two kinds of avoiding strategies;
- sustainability themed (investment focused in sustainability projects), best in class (investment in companies with higher ESG performance) and ESG integration (incorporate ESG in decision-making process) as three strategies for promoting sustainability;
- impact investment (investments aimed at social/environmental benefit) and engagement voting (active investor role in companies) as two strategies for changing and adapting to sustainability.

Paying attention to the institutional aspect of ESG risk management, regulations, and institutions established to manage macro- and microprudential risk in the banking sector after the 2008 crisis played an important role in the scope of banks' management. In this context, the assumptions of the banking union and fiscal union are important. Environmental risk in the context of its impact on the stability of the banking sector is included in paragraph 510 of Basel II and Basel III (Pillar 1), which obliges banks to adequately monitor environmental risk in relation to legal security for repayment. Paragraph 510 deals with transactions that may be at environmental risk by providing bank financing to a borrower that is subject to environmental risk, which determines its creditworthiness and thus its ability to repay the debt and the determination of the value of its legal collateral.

The literature on the subject emphasizes that the role of ESG risk and its impact on the financial results of business entities will grow due to the growing importance of factors and aspects in the activity such as

- intensification of interactions and relationships between climate change and biodiversity and modern technologies, which will determine a new approach to investment, risk, and rates of return;
- high ESG ratings, which will significantly determine the cost of obtaining capital and in particular, the low cost of obtaining debt financing, which may constitute significant competition and a threat to financing in the form of municipal bonds, the cost of which may be higher and limit the access of self-employed local government to this source of funding;
- ESG risk-weighted asset classes that will be reclassified (assets not subject to environmental risk as a result of changes in the environment may increase exposure to this risk), which will allow for the development of new strategies for diversification and transaction risk management.

7.3 Sustainable Banks' Business Models

The influence of ESG factors applies in particular to banks that play a key role in ensuring financial stability and security, which will simultaneously condition the processes of economic development. The role of banks as financial intermediaries means that the challenges created by ESG factors for the financial system require them to implement adjustment measures. The scope of these activities includes the development and implementation of ESG risk management systems, methodologies, and procedures; the adjustment of the product and service offer to ensure support and promotion of good practices in the field of activities for sustainable development; and, finally, the implementation of the business model. The scope of the activities is based on creating sustainable value (Stampe, 2014). Financial institutions respond to ESG risk in different ways; to a large extent, this response depends on the type of financial institution (investment banks react differently from commercial banks). The common response, however, is to minimize ESG risk through appropriate shaping of the product offer and through changes in banks' business models.

The banking sector, as an integral part of the financial market, is undergoing changes to adapt to financing sustainable development; however, both the method of transformation and adaptation of this sector as well as the conceptual apparatus accompanying the changes remain heterogeneous. The concept of values-based banking presents a comprehensive approach to integrating social and environmental values into the mainstream of banking and its considerations. It is a banking model based on (1) the triple bottom-line concept; (2) the servant role of banking in relation to society and the economy; (3) building long-term relationships with stakeholders; (4) a long-term, self-supporting, and non-influencing approach; and (5) inclusive and transparent governance. Banking based on values is characterized by a much smaller scale of activity than traditional banking, the active participation of the public in the discussion on financial needs, and a model of institutions that should meet these needs. Banking based on values is a concept around which it is possible to build new socio-economic institutional solutions focused on the development of an economy based on social and environmental standards that fit in with the postulates of sustainable development (Values, 2015). Another example is the concept of sustainable banking. Sustainable banking is a term used to describe financial services and products that have been designed to meet social needs and protect the environment while ensuring the operating profit of sustainable banks (Yip & Bocken, 2018).

The concept of sustainable banking is derived from ethical banking or, in the narrower approach, from the ethical policy of banks, which indicate the standards and activities in the field of whether or not to invest in specific funds, or cooperating or withdrawing from cooperation with specific entities. It is worth paying attention to the interpenetration of the boundaries of the pillars of sustainable development in green banking. Green banking indirectly interacts with the social pillar of sustainable development because the quality of the natural environment, which is the focus of green banking, determines the quality of human life and the incidence of certain diseases, the incidence of which increases with increasing environmental pollution (Yip & Bocken, 2018).

Increasing the effectiveness of banks' operations for sustainable development requires these institutions to adjust their business models so that they are based on building sustainable value (Fig. 7.4). Converting traditional business models to sustainable models is a complex process for which banks must prepare. In the transformation of business models, the triple bottom line (TPL) approach is used, consisting of three components: people, planet, and profit. The business model is usually described by the components that compose the business model canvas, which consists of the analysis of areas such as key partners, key activities, value, customer relations, customers, key resources, methods of reaching, financial flows, and cost structure.

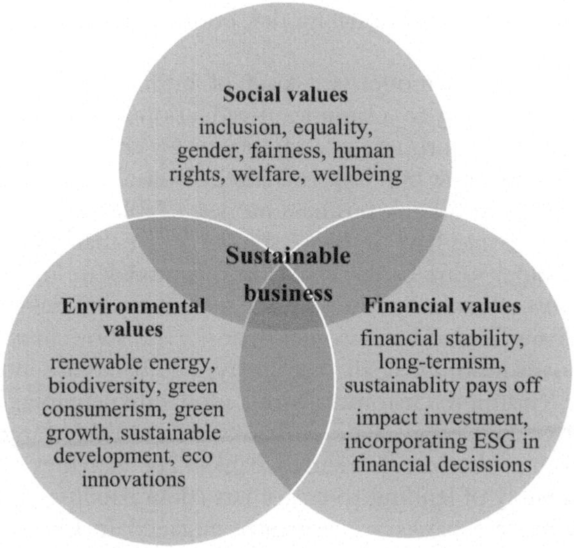

Fig. 7.4 Sustainable business (*Source* Own elaboration based on https://fuerteventuratimes.com/business/why-sustainable-businesses-are-more-profitable-how-to-be-one/ [access 19.02.2021])

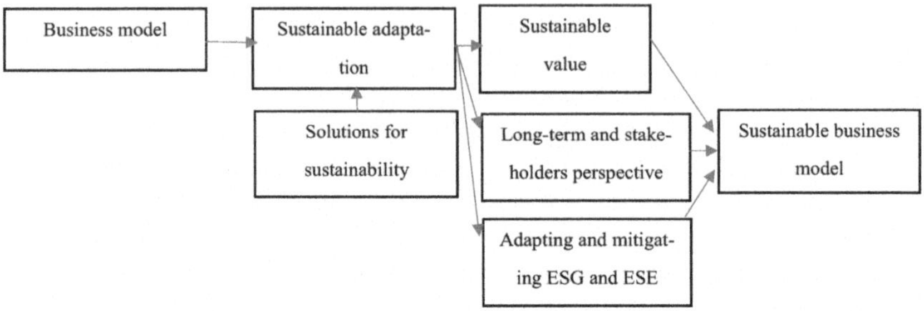

Fig. 7.5 Sustainable business model evolution (*Source* Own elaboration based on M. Geissdoerfer, S. Morioka, M. Carvalho, & S. Evans. (2018). Business models and supply chains for the circular economy. *Journal of Cleaner Production*, 190. https://doi.org/10.1016/j.jclepro.2018.04.159)

The sustainable business model, unlike the traditional model, considers the components of the traditional model, but from stakeholders and long-term perspective (Fig. 7.5).

Sustainable business model is a kind of business model that creates benefits by contributing to a significant reduction of the negative impact of business on the environment and society by changing the organization's approach to value building (Schaltegger et al., 2016). Ayadi et al. (2019) distinguished 5 bank business models (Table 7.3).

Focused retail banks have an ownership structure that is slightly skewed toward stakeholder value banks. Banks in this model are mainly cooperative banks. Most of banks in this models based on traditional services, such as customer loans funded by customer deposits and show a moderate non-performing loans ratio. Diversified retail (type 1) banks are modest in size and combine lending to customers with a moderate percentage of trading activities. Diversified retail (type 2) banks are relatively large in size as compared to the other retail-oriented banks. The activities of banks this type consist mainly of lending to customers (debt liabilities and customer deposits). Wholesale banks are primarily engaged in interbank lending and borrowing and are primarily classified as shareholder value banks. Investment banks are involved mainly in trading activities (Ayadi et al., 2019). Some studies use the level of sustainability to assess and measure the sustainable bank business models. According to this approach sustainable bank business model might be: red, blue, orange, green, yellow, or turquoise (Table 7.4).

Table 7.3 Bank business models according to Ayadi et al. (2019)

Model/criteria	*Model 1—focussed retail*	*Model 2—diversified Retail (Type 1)*	*Model 3—diversified retail (Type 2)*	*Model 4—wholesale*	*Model 5—investment*
Ownership	Cooperative, savings banks	Small, cooperatives, savings and public	Cooperative and savings banks	Predominantly commercial banks, but largest share of public banks	Predominantly commercial banks, but substantial share of cooperative banks
Financial performance and operational efficiency	Relative high returns except for crises	Returns as well as operational efficiency deteriorated during the crises	Returns most stable and only model not posting a loss in a single year	Returns stable, although the reduction during crisis	Returns rather stable, except for crisis
Contribution to the real economy	High stable customer loan growth	High customer loan growth (but lower than Focused retail during and after the crises)	Low customer loan growth during financial crisis and negative during/after economic crisis	Relatively high customer loan growth during crises but with some years with negative sign	The loan growth deteriorated relatively more during the financial and economic crises
Risk	The second highest distance to default	The second highest distance to default	Low distance to default	Low distance to default	The lowest distance to default
Response to regulation and resolution	Moderate Tier-1 and high tangible equity; low bail-in contribution. The second highest MREL among BM	The highest Tier-1 among retail-oriented banks; moderate tangible equity; moderate bail-in contribution and the highest MREL	The lowest Tier-1 and low tangible equity low bail-in contribution least liquidity Moderate level of MREL	Sharply increasing Tier- 1 and tangible equity high bail-in contribution. Moderate MREL	High Tier-1 and the lowest tangible equity; the highest bail-in contribution Lowest MREL

Source Own elaboration based on R. Ayadi, D. Cucinelli, & W. P. de Groen. (2019). *Banking business model monitor Europe: Performance, risk, response to regulation and resolution 2005–2017* (pp. 105–107). London: Cass Business School, City University

Table 7.4 The level of sustainability in banks business models

Model	*Description*	*ESG risk*
Red—pre-Corporate Sustainability (CS)	There is basically no ambition for CS, CS might be external driven	Not incorporated
Blue—Compliance-driven CS	CS seems to be a kind of duty and obligation	Incorporated partially
Orange—Profit-driven CS	CS is promoted if profitable	Incorporated partially
Green—Caring CS	CS initiatives go beyond legal compliance and beyond profit considerations	Incorporated partially
Yellow—Synergistic CS	Sustainability is crucial	Incorporated
Turquoise—Holistic CS	Sustainability is the only way	Incorporated

Source Own elaboration based on W. Schutten. (2016). ***Sustainable banking business models***. Research Report. Master Business Administration, University of Twente, pp. 105–107

Bocken et al. (2014) indicated the basic archetypes of sustainable business models (Bocken et al., 2014, pp. 42–56) (Table 7.5).

- technological (the use of renewable sources, digitization, and pro-environmental activities),
- social (pro-social activities, sponsorship, microfinance, and inclusive activities on the labor market), and
- organizational (promotional, risk management, and sustainable products and services).

Value is at the heart of any business model. This is indicated by the ranking of the components of banks' business models prepared by Nosratabadi et al. (2020). According to the ranking, the most important elements of business models include value proposition, core competencies, financial aspects, business processes, target customers, resources, technology, customer interface, and partner networks (Nosratabadi et al., 2020, p. 11). Value-building occurs in various ways, and value varies depending on the archetype the business model represents and the kind of value being analyzed. With reference to the archetypes presented in Table 7.5, the value in the sustainable business models of banks can be analyzed from the perspective of the value proposition and how it is built and delivered (value creation and delivery); or the methods of its disclosure (value capture). Gilding et al. (2002) proposed four key drivers of value creation:

Table 7.5 The basic archetypes of sustainable business models

Type of archetype	*Key assumptions*	*Practical aspect*
Technological	1. Maximizing material and energy efficiency 2. Replacing material activities with digitization processes	• Electronic documents in place of paper ones • Online trading, e-banking, ATM, mobile applications, electronic payments
Social	3. Encouraging to be effective and efficient 4. Adjusting management practices toward sustainable management 5. Building value on an inclusive approach	• Needs-based selling, promoting reasonable indebtedness, improving the product offer, evaluating the product quality–price relationship • Re-employing retired workers; sponsorship of companies implementing CSR, debit cards for deaf and blind employees, refusing to cooperate with companies violating environmental and social standards, giving preference to charity donations instead of flowers • Supporting clients in a difficult situation (including natural disasters), microfinance, loans for SMEs, supporting the first house purchases
Organizational	6. Reorientation of goals toward social and environmental 7. Restrictive creditworthiness assessment 8. Sustainable financial products	• Lend/finance only "sustainable" companies • Selective micro-financing only for the "poorest" • E&S (environmental and social) rating according to Standard Chartered (negative screening) • Equatorial principles (mainly for project financing) • Green bonds, sustainable shipment letter of credit, SRI, preferential interest rate when buying machinery in installments with a positive impact on the environment/energy efficiency

Source Own elaboration based on A. W. H. Yip & N. M. P. Bocken: *Sustainable*, op. cit. p. 163

margin improvement, risk reduction, growth enhancement, and capital efficiency (Stankeviciene & Nikonorova, 2014). A detailed approach to value analysis according to the archetypes of sustainable business models is presented in Table 7.6.

Building sustainable value in bank business models, as shown in Table 7.6, is based on environmental, social, or governance factors. Therefore, in the context of the typology of bank business models, there are green business models, social models, governance models, or the mixed/hybrid models using a mix of ESG factors to build sustainable value in banks 'business models. Zhelyazkova and Kitanov (2015, p. 312) referred to the green bank business model, "The application of a green bank business model ensures that the bank is making profit by encouraging environmentally responsible behavior" (Zhelyazkova & Kitanov, 2015, p. 313). The aim of a bank's green business model is therefore to stimulate and support environmentally friendly behavior of banks through their basic activities. The green business model is based on the distribution channels of services that do not generate carbon footprints, and involved green financial products, bank investments in green companies and funds, green financial reporting, the green office model, and segmentation of clients based on the type of business (preferred business partners and environmentally friendly clients).

Business models based on the social component assume that employee rights are respected, engaging banks in supporting initiatives to defend human rights, products and services supporting social inclusion, and participation in charity actions and initiatives. An important area of activity is supporting socially responsible investments and business on the part of business partners and customers. In this approach, the business models of social banks play an important role (Cornée et al., 2019). Cornée et al. defined social banks as "as social enterprises that run banking activities with the social mission of supplying credit to other social enterprises, which are typically less profitable than for-profit businesses" (Cornée et al., 2019, p. 1). A key element of the social banks' business model is microfinance as an instrument to combat poverty and social exclusion (Cornée et al., 2019, p. 32).

The last type of sustainable business models based on the governance component adopts the assumption of building value in the bank's business model, which is based on transparency, promoting ethics in business, and broadly understood good corporate governance practices.

Table 7.6 Value building in sustainable banks' business models

Type of archetype	*Value*	*Value creation*	*Benefits*	*Value added comparing to conventional models*
1. Maximizing material and energy efficiency	Services that consume less energy generate less waste with similar functionality	Focus on innovation in internal processes	Costs are reduced by increasing efficiency, while increasing profitability	Comprehensive efficiency improvement strategies are replaced by a fragmentary approach
2. Replacing material activities with digitization processes	Reducing the negative impact on the environment by using electronic means/carriers in business activities	Innovations in the process of service delivery ensuring shortening of the delivery time, cost savings and adequacy of distribution channels to customer expectations	Increase in revenues created by the possibility of increasing the frequency of electronic services, a decrease in costs related to the reduction of costs generated by the human factor	Digitization provides banks with innovation, easy and quick contact with the customer without the need to maintain traditional bank branches
3. Encouraging to be effective and efficient	Solutions to reduce overstated demand through a revised, more restrictive approach to customer selection to eliminate unfair practices and moral hazard. Focus on building relationships with the client	The activities may concern changes in the composition of the front office staff, changes in the remuneration structure (the relation between the fixed part, the variable part depending on the effects of work), increasing the sales corresponding to the demand	Customer satisfaction and loyalty can increase and provide more business revenue. The compliance risk is lower and reduces the likelihood of penalties imposed by market regulators	Banks are moving away from selling based on the "sell more" principle and focusing on selling the best products and services most suited to the needs of customers

(continued)

Table 7.6 (continued)

Type of archetype	*Value*	*Value creation*	*Benefits*	*Value added comparing to conventional models*
4. Adjusting management practices toward sustainable management	Building long-term relationships with stakeholders as an important element of building value	Making sure that all actions and activities are aimed at ensuring stakeholder satisfaction. The value chain is geared toward social and environmental benefits	Strategies are focused on building brand value, savings and securing the business in the future. Stakeholder satisfaction provides multi-directional long-term benefits	Compared to traditional CSR, banks try to integrate the approach to value built with stakeholder satisfaction into their core operations and leverage
5. Building value on an inclusive approach	Inclusion takes place with the use of product innovations that allow you to serve customers who have not been served so far at too high transaction risk	Innovations are a key element of risk reduction using credit scoring, portfolio analysis and SME risk management	Increasing market share, more business opportunities resulting from customer loyalty and the expectation of an increase in profitability in the future, when a given group of customers becomes more profitable, social benefits are also included	Banks engage in transactions with disadvantaged clients, counting on the improvement of their situation in the future and an increase in profitability on these relationships in the long term
6. Reorientation of goals toward social and environmental	Creating social and environmental benefits by addressing the product offer dedicated to customer needs	The sustainable business criterion is the customer selection criterion; banks specialize in servicing sustainable business and count on economies of scale	Providing banking services only to sustainable business and disadvantaged clients based on positive screening and not only on environmental and social benchmarks	Banks dedicate their services to segments of sustainable business clients

(continued)

Table 7.6 (continued)

Type of archetype	*Value*	*Value creation*	*Benefits*	*Value added comparing to conventional models*
7. Restrictive creditworthiness assessment	Financial factors are incorporated in credit and loan decisions allowing for negative screening of projects that do not meet the criteria compatible with sustainable development	The ESG risk is lower, thus the cost of obtaining capital is lower and the risk of the so-called Bad debts	Lending dedicated only to enterprises with zero or low ESG risk reduces the availability of capital resources for entities that do not meet sustainable criteria	Banks conduct risk and transaction analysis in a more comprehensive way, taking into account ESG risk and thus eliminating customers who do not meet the criteria of sustainable business
8. Sustainable financial products	On the assets and liabilities side, products dedicated to savers, investors and borrowers are created	Product innovations open and develop the market of sustainable finance, supporting sustainable development	Product innovation enables participation in sustainable development that can provide a platform for both savers and borrowers to obtain a financial return on a sustainable business	Banks focus on the new market demand for sustainable business by providing clients with new platforms and financial products

Source A. W. H. Yip & N. M. P. Bocken: *Sustainable*, op. cit. p. 163

Each of the three types of bank sustainable business models incorporates a specific ESG risk component. Table 7.7 shows the relationship between ESG risk and the types of sustainable business models based on the ESG components.

The sustainable business models of banks, unlike conventional business models, differ in the factors on which these models are based. In the case

Table 7.7 The relationship between sustainable banks business models and ESG risk

Degree	*Types of action*	*Type of risk addressed by sustainable business model*
Degree 1 ESG Minimal approach	• The bank minimally manages the ESG risk in its operating activities (e.g., carbon footprint, water management, philanthropic initiatives) • The Bank publishes reports on basic initiatives in ESG risk management • The bank is not aware of the real impact of ESG risk on operating activities	No one
Degree 2 ESG aware approach	• In addition to managing the direct impact of ESG risk, the bank also undertakes a comprehensive attempt to manage ESG risk in its operating activities • The bank is aware of the impact of ESG risk on its business • The bank starts incorporating ESG risk into the risk management process and the process of developing banking products, including by "greening" them or offering products dedicated to social and environmental issues • Opportunistic management is based on "must have" and the so-called "Low hanging fruits"	Mainly environmental

(continued)

Table 7.7 (continued)

Degree	*Types of action*	*Type of risk addressed by sustainable business model*
Degree 3 ESG systematic approach	• The bank begins to systematically integrate ESG risk into the risk management process, including by developing policies dedicated to high-risk sectors and integrating these policies with the risk management process • The bank's awareness of the role and significance of ESG risk in operating activities is growing • ESG risk control becomes a permanent element of the activity and risk management process • This approach is mainly based on management risk • Reporting is extended to include information on ESG risk targets and their implementation as well as statistics	Environmental, Social, Governance
Degree 4 ESG complex approach	• The bank fully incorporates ESG risk in its activities and decision-making processes • The bank begins to perceive ESG risk as a strategic element of business management • The Bank perceives ESG as a key element of risk management and a component of building a competitive advantage (including building customer relationships) • Annual ESG reporting dedicated to stakeholders	Environmental, Social, Governance

(continued)

Table 7.7 (continued)

Degree	*Types of action*	*Type of risk addressed by sustainable business model*
Degree 5 ESG as dominant approach	• The bank treats ESG as the dominant strategic issue and an element determining innovation • ESG remains integrated with the valuation and asset management processes as well as the pricing policy to maximize market impact • Business processes are being reoriented toward ESG and new business models are emerging • Banks engage in social issues, including building a low-carbon economy • ESG reporting becomes part of daily communication	Environmental, Social, Governance

Source Own elaboration based on ***Environmental, social and governance integration for banks: A guide to starting implementation***. Ed. Jeanne Stampe, World Wide Fund for Nature (Formerly World Wildlife Fund), Gland, Switzerland, 2014, p. 25

of traditional models, these are economic factors; in the case of sustainable business models, these are non-financial factors. The knowledge and experience in shaping sustainable business models are limited compared to traditional business models.

7.4 Conclusion

In the post-2008 crisis period, banks have faced a number of challenges: first, challenges related to the adjustment of the risk management system to consider macroprudential risk and ensure the stability and security of the banking sector on a global scale. The growing role of ESG risk and its impact on the situation of the banking sector have also forced the implementation of adjustment and adaptation measures. Adapting the operations of banks to the sustainability requirements necessitates changes in strategic management and, consequently, changes in the banks' business models.

The COVID-19 pandemic accelerated the building of business models by banks that include building sustainable value. The pandemic-driven changes have mainly affected the channels of financial services. Electronic distribution channels for banking services in the pandemic era have become common. Therefore, the pandemic has accelerated actions aimed at implementing the green office, which involves digitizing service delivery processes while reducing office paper consumption. In addition, it assumes a reduction in electricity consumption, which, in the case of digitization of office processes, occurs when the functioning of the office is based on renewable energy sources.

The strongly influential factor in this respect is primarily the environmental aspect. Environmental risk is a burdening banks both in the context of loss of assets securing loans, but also the financial condition of banks in terms of credit receivables from the bank's customers exposed to this risk. In this context, banks may act directly or indirectly. Direct action consists of implementing eco-innovation in the operational activities of banks, and indirect action is related to the influence of banks on their clients' business models by encouraging them to introduce eco-innovations, e.g., by offering more favorable terms of the transaction (mainly the cost of capital).

In the coming years, a transformation of banks' business models toward more sustainable ones should be expected due to the growing role and impact of non-financial factors on the operating activities of banks. Activities in the field of building sustainable value in the business model will focus on the product offer (sustainable financial products), directions of capital allocation by banks in projects supporting sustainable development, respecting employee rights by building employment policy friendly to employees with a wide range of employee benefits, increasing the role of technology in providing services and thus limiting energy consumption, supporting charity campaigns, changing the procedures of internal banks toward comprehensive risk and ESG management, and including ESG risk in the bank's short-, medium-, and long-term decisions.

References

Ayadi, R., Cucinelli, D., & de Groen, W. P. (2019). *The banking business models (BBM) monitor 2019 Europe: Performance, risk, response to regulation and resolution: 2005–2017* (pp. 65–101). London: Cass Business School.

Bank Track. (2006). https://www.banktrack.org/news/the_dos_and_donts_of_sustainable_banking. Accessed 12 Nov 2020.

Bocken, N. M. P., Short, S. W., Rana, P., & Evans, S. (2014). A literature and practice review to develop sustainable business model archetypes. *Journal of Cleaner Production, 65*, s.42–56.

Cornée, S., Kalmi, P., & Szafarz, A. (2019). *The business model of social banks*. Kyklos. 73. https://doi.org/10.1111/kykl.12221. https://www.researchgate.net/publication/338096801_The_Business_Model_of_Social_Banks. Accessed 16 Nov 2020.

Escrig-Olmedo, E., Ángeles Fernández-Izquierdo, M., Ferrero-Ferrero, I., Rivera-Lirio, J. M., & Muñoz-Torres, M. J. (2019). Rating the raters: Evaluating how ESG rating agencies integrate sustainability principles. *Sustainability, 11,* 915.

Fullwiler, S. T. (2015). *Sustainable finance: Building a more general theory of finance* (Working Paper No. 106).

Gilding, P., Hogarth, M., & Don Reed, D. (2002). *Single bottom line sustainability: How a values centred approach to corporate sustainability can pay off for shareholders and society* (pp. 18–19). Sydney: Australia: Ecos Corporation.

https://www.bankinghub.eu/banking/research-markets/esg-investments-current-trends

Nosratabadi, S., Pinter, G., Mosavi, A., & Semperger, S. (2020). Sustainable banking: Evaluation of the European business models. *Sustainability, 12,* 2314.

Schaltegger, S., Hansen, E., & Lüdeke-Freund, F. (2016). Business models for sustainability: Origins, present research, and future avenues. *Organization and Environment, 29,* 3–10. https://doi.org/10.1177/1086026615599806.

Schutten W. (2016). *Sustainable banking business models: Research report*. Master Business Administration, University of Twente.

Stampe, J. (Ed.). (2014). *Environmental, social and governance integration for banks: A guide to starting implementation*. World Wide Fund for Nature (Formerly World Wildlife Fund). Gland, Switzerland.

Stankeviciene, J., & Nikonorova, M., (2014) Sustainable value creation in commercial banks during financial crisis, contemporary issues in business, management and education 2013. *Procedia—Social and Behavioral Sciences, 110,* 1197–1208.

Values based banking. (2015). *Bringing the voice of the citizen into finance* (UNEP Inquiry, IWP 15/03, p. 6). https://unepinquiry.org/wp-content/uploads/2015/04/Values_Based_Banking.pdf. Accessed 5 Jan 2020.

What Is ISO 26000—Guidance on Social Responsibility? https://asq.org/quality-resources/iso-26000.

Yip, A. W. H., & Bocken, N. M. P. (2018). Sustainable business model archetypes for the banking industry. *Journal of Cleaner Production, 174,* 150–169.

Zhelyazkova, V., & Kitanov, Y. (2015). Green banking-definition, scope and proposed business model. *Ecology and Safety*. Journal of International Scientific Publications, 9. https://www.researchgate.net/publication/330202134_GREEN_BANKING-DEFINITION_SCOPE_AND_PROPOSED_BUSINESS_MODEL/citation/download. Accessed 16 Nov 2020.

Ziolo, M., Filipiak, B. Z., Bąk, I., & Cheba, K. (2019). How to design more sustainable financial systems: The roles of environmental, social, and governance factors in the decision-making process. *Sustainability, 11,* 5604.

CHAPTER 8

Relationships Between Banks' and Companies' Sustainable Business Models

What does a change in the business model of a financial institution—a bank mean in the practice of enterprises? Why is it important for companies to take responsibility for their actions, especially today? Why is this responsibility and pro-environmental behavior of enterprises becoming an important decision-making aspect for the bank? These are the questions the chapter responses and the issues that determine the spheres of influence bank–enterprise in sustainable business models.

8.1 Spheres of Influence of Banks on Companies

What does a change in the business model of a financial institution (a bank) mean in the practice of enterprises? Why is it important for companies to take responsibility for their actions, especially today? Why is this responsibility and pro-environmental behavior of enterprises becoming an important decision-making aspect for banks? These are the issues that determine the spheres of influence banks and enterprises in sustainable business models. Businesses, and in particular enterprises, are forced to take care of their immediate surroundings and the environment in which they function. They are forced by changes in stakeholders or changes in the law, and by the feeling of running out of resources (the economic aspect, the aspect of responsibility, and the aspect of matching the market gap). Companies are also required to act for the benefit of society. For

© The Author(s), under exclusive license to Springer Nature Switzerland AG 2021

M. Ziolo et al., *Sustainability in Bank and Corporate Business Models*, Palgrave Studies in Impact Finance,
https://doi.org/10.1007/978-3-030-72098-8_8

the company to develop in a sustainable manner—which is what every company operating toward sustainability strives for—the people responsible for it must be aware of the relationships that exist between its activities and changes that occur in the environment or society. These changes are also forced by banks that also belong to the business group and are profit-oriented.

Actions for sustainable development must be long-term actions for two reasons. Firstly, one-off actions do not bring the intended results or benefits. Long-term operation allows for increasing brand awareness, building a community of loyal customers around it who will appreciate additional activities, which is a side effect of investing in sustainable development. Secondly, enterprise does not have to take such steps alone. It is looking for business partners (banks) who will financially support the undertaken activities, but also often share the same environmental ideas, i.e., their business models also consider the SDGs.

As business models will be adapted to the SDGs, as business advantage is increasingly experienced in sustainability, both banks and their customers, and in particular enterprises, will seek common elements, or even common ground of understanding. Thus, the following spheres of influence of banks on corporations should be seen from the viewpoint of changing and adjusting business models that result from canvas:

1. business partners, including supplier policy;
2. service and goods distribution channels;
3. key activities/spheres of activity;
4. customer segments; and
5. key/strategic resources.

The indicated spheres of influence of banks on companies constitute this area, which becomes a kind of influence of the banks on enterprises and enterprises on banks in the area of changes in the business model, in terms of sustainability. Mutual interactions between the bank and the enterprise in terms of the impact on business models are presented in Fig. 8.1.

Figure 8.1 indicates that both the bank and its client (the enterprise) may influence the partner's business models under the influence of various factors related to sustainability. It may be a change forced by environmental factors, changes in mentality and social responsibility, ESG risk,

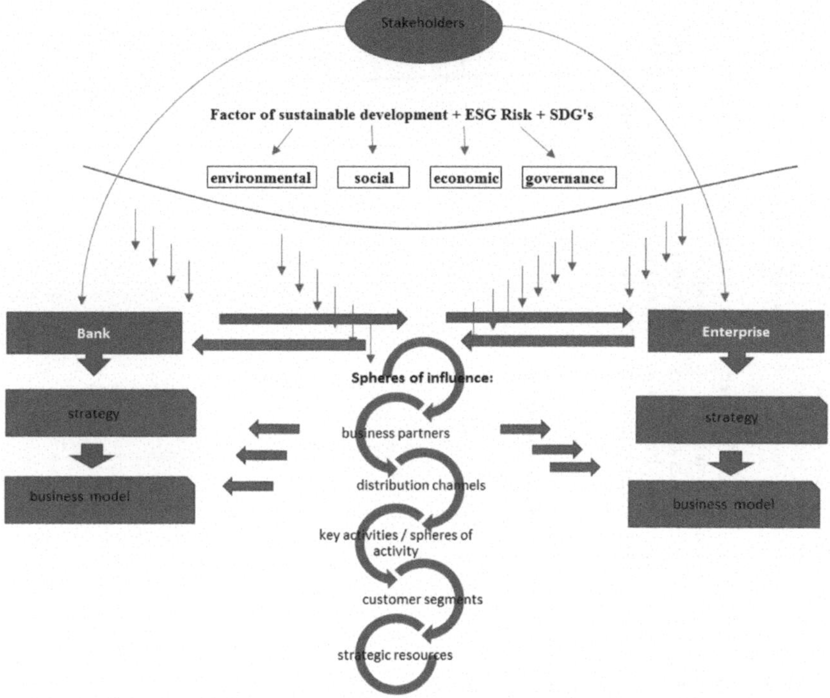

Fig. 8.1 Mutual interactions between the bank and the enterprise in terms of the impact on business models (*Source* Own elaboration)

or compliance with the SDGs (for various reasons). Changes in business models may be forced by the bank on its clients because the bank, by changing its business model, will change distribution channels, key activities, or decide to move toward green market segments. Remember that (as discussed in Chapter 2) the choice of a bank's business model is influenced by a number of exogenous and endogenous factors. Risk is also considered, including ESG risk. Table 8.1 indicates the interactions caused by the type of factor changing the bank's business model, which may imply changes in business models toward companies.

Enterprises will increasingly look for instruments and green products in which they could invest, indicating that this green wave is coming. The elements of green finance involve and require more transparency and

Table 8.1 The interactions caused by the type of factor changing the bank's business model

Spheres of influence of banks	*Characteristics of the impact*	*Type of key factor*
Customer segments	The bank will define the recipient of the offer or entire target groups. Segmentation makes it possible to identify the characteristics of buyers of banking services and their needs, but also to stimulate their actions toward sustainability. This in turn defines the product strategy and directional (nodal) changes. They can be adaptive or radical The bank has to calculate the change because it is the customers who are the source of its income	Exogenous: political and legal; social, economic Endogenous: financial resources, organizational resources, innovative resources, market resources

Spheres of influence of banks	*Characteristics of the impact*	*Type of key factor*
Business partners, including supplier policy	In this sphere of the bank's influence on the enterprise toward changing the business model containing the elements of sustainability desired by the bank, there are activities aimed at analyzing the network of suppliers, contractors, producers and all contractors on which the company's operation depends Strategic partnerships with competitors operating according to the principles of sustainability can also be considered. Motivation to establish partnerships with entities operating according to the principles of sustainability, and in particular to combine activities for sustainability with the desire to reduce costs, minimize risk or produce a better product. The results of the analysis will influence the bank's decisions regarding financing of current transactions or the impact on pro-ecological investments The bank will analyze whether the company's cooperation with its business partners who maintain sustainability standards contributes to the optimization and economy of scale of the client's business. It is also worth indicating the scope of interactions with clients—from personal to automated ones, and define what and how will be communicated to the client in terms of changes regarding the introduction of environmental and CSR goals in the bank	Exogenous: social, economic Endogenous: financial resources, organizational resources, innovative resources, market resources, human resources

(continued)

Table 8.1 (continued)

Spheres of influence of banks	*Characteristics of the impact*	*Type of key factor*
Key/strategic resources	On the one hand, resources make it possible to create a product, reach the customer, support and motivate the customer to change toward sustainability. This important resource belongs to the group of intellectual resources, and is associated with the possession of staff (human capital) who feels responsible for acting in accordance with the idea of sustainability. Intellectual resources also include e.g. patents, copyrights or customer data managed and stored in accordance with the CRS concept Physical (human) resources are also important, especially those that contribute to efficient customer service, especially IT systems. The resources also include the capital held (and above all the possibility of generating them), which are the basis of the bank's operations. Showing that the bank operates in accordance with the principles of CSR prevents infringement of employee rights	Exogenous: social, economic Endogenous: financial resources, organizational resources, innovative resources, market resources, human resources

Spheres of influence of banks	*Characteristics of the impact*	*Type of key factor*
Service and goods distribution channels	Channels are a way or a place to provide information to you about the products and services our company provides. In addition, of course, they are a method of purchasing a product and a place for later evaluation of cooperation and expressing opinion on the offer itself. Communication, distribution and sales are three issues that constitute the key elements at the interface between the enterprise and the bank. The customer will be informed about the offer and sustainability goals through distribution channels. In addition, the bank's decisions are communicated and information on the necessity to adapt to the bank's requirements is distributed. Correct definition of distribution channels additionally plays an important role in creating experience customers and their reaction to adjustments toward sustainability	Exogenous: social, economic Endogenous: financial resources, organizational resources, innovative resources, human resources

(continued)

Table 8.1 (continued)

Spheres of influence of banks	*Characteristics of the impact*	*Type of key factor*
Key activities/spheres of activity	This element is the most important task that must be performed toward the implementation of the idea of sustainability and the correct creation of value propositions, contact with customers and most importantly—generating revenues. Basic forms the activities carried out by the bank should be described in such a way as to indicate those requiring immediate adjustment of customers to the goals related to sustainability, define the achievement of which will be spread over time, and indicate which spheres of activity will not be related to the implementation of the sustainability goals. The bank must indicate not only the spheres of activity in connection with sustainability, but also focus on the quality of activities—an appropriate process that will improve work. This element is important for customers–enterprises. By the company's spheres of activity, customers recognize what kinds of SDG's goals are supported by the bank through products and services, and what kinds of activities for social responsibility will be represented	Exogenous: political and legal; social, economic, technological Endogenous: financial resources, organizational resources, innovative resources, market resources, human resources

Source Own elaboration based on: Global4 (2020), Klimontowicz & Pyka (2016), Raport (2020) more information from and to the client and the investor than is the case with traditional instruments. Accordingly, stakeholders, including banks, will require increasing amounts of information from the supervised entities regarding how they deal with climate risks on the capital or money market. Both banks and enterprises are realizing that environmental regulations will continue to tighten. So, they need to adjust not only their accounting or systems, they also have to adapt to GRI standards and to future challenges related to reporting so that it is consistent with future guidelines. This means that to raise capital, entrepreneurs will have to provide relevant information so that the financing and ESG risks can be assessed. Banks will have to adapt their procedures to assess not only funding risk (credit risk) but also ESG risk and other environmental risks that have a financial dimension. This is an important point that indicates changes in business strategies (Raport, 2020).

Exogenous factors of an economic (the level of development of the country, changes in prices on the markets, and economic situation), political and legal (monetary and fiscal policy, legal regulations, recommendations, and directives), social (demography, changes in customer needs and expectations, and lifestyle changes), and technological (development of telecommunications, mobile, and business support technologies) nature are important motived for making changes to business models, but as already indicated, this motive is caused by external forces or the desire to adapt related to a specific effect and goal. Endogenous factors are a derivative of the strategic resources (tangible, financial, and human) held by the bank and result from the operational and strategic decisions made. Changes in banks' business models toward sustainability will necessitate changes in enterprises. However, the bank, changing toward sustainability, will force changes on customers. Remember that changes in banks' business models toward sustainability will reduce the scale of the bank's operations while increasing safety and accountability, which will lead to smaller but more stable profits (KPMG, 2011, p. 8). This may mean limiting the fields of cooperation for the benefit of quality and safety for the company, and it may lead to changes in the cost structure and bank charges, which can affect the cost canvas of banks and enterprises.

Changing the business model forces a change in key areas of activity, and the introduction of SDGs or a partial redirection toward sustainability will result in narrowing the key areas of activity, which will lead to greater specialization and withdrawal from areas that are not related to the implementation of the defined sustainability and CSR goals. Due to this approach, the bank may lose customers or it will be forced to redefine the areas of its activity, adapting to new customers in the area of sustainability, whose activities force the definition of new specializations corresponding to their needs related to the implementation of their own goals in line with the SDGs.

The redefinition will also include withdrawing from the concept of universal banking offering a full range of banking services toward solely sustainability banking; thus, it will be easier to identify common spheres of influence of the bank and the client (the enterprise). This shift, in turn, causes the loss of certain markets, but forces focus on fewer but fast-developing markets. Thus, sustainability in business models is a specific promise of increasing the market share of sustainability services, and resignation from traditional non-green markets.

8.2 The Common Framework for Banks' and Companies' Business Models

Banks and enterprises interact with each other in the economy. Banks act as business partners of enterprises, and enterprises use banking services and products, including consulting services. Banks also play a special role as public trust institutions. Banks define the terms of cooperation with enterprises depending on their profile and segment. Depending on the type of transaction and the risk, these conditions may be more or less restrictive. In the context of sustainability, it is important to examine the bank as an institution that influences or can influence the enterprise so that it adjusts its model toward a sustainable business model, i.e., a model that also implements sustainable banking.

A common field of influence for banks and enterprises is the business model affecting the financial situation of both. An expression of the decisions made by the company is its financial situation, which is reflected in the financial condition. The company's financial situation is assessed using a number of indicators informing about the state of liquidity, debt, profitability, operational efficiency, etc. Each of these aspects reflects financial decisions made by decision-makers, e.g., in the field of business models, sources of financing, management of current assets, investment policy, tax strategy, dividend policy, margin, and/or risk. When treating financial analysis (in particular, ratio analysis) as a source of information about an enterprise, financial indicators are often classified considering the recipient of information. In this context, groups of indicators are indicated that are of particular informative importance from the point of view of three groups of entities: managers, owners, and lenders.

The broadly understood profitability analysis consists of analyzing the level of activity and the nature of the financial result; in particular, as a result of the analysis, whether the enterprise generates profit on operating activities, and how the analysis and dynamics of revenues and costs are shaped should be established. Further determinations should concern the extent to which the result at each level is burdened with tax and interest costs, the state of the depreciation of the property, and how its depreciation progresses. From the cost management point of view, it is crucial to define the relationship between fixed and variable costs and to indicate the dynamics of changes in each group. It is important to identify and skillfully diagnose cost sources and cost centers and determine their impact on the company's financial situation, including the financial result.

The decisions of enterprises affecting the indicated parameters of the financial situation indicate the tax strategy chosen by the enterprise, the position and negotiating power of the enterprise in relation to capital donors, and, consequently, the cost of capital. The size of the enterprise, its competitive advantage, and market position determine the company's ability to generate profit and financial flows. A direct relationship exists between a company's financial decisions and its financial situation, as described by a number of financial indicators. This applies to both current and strategic decisions regarding, among others: profitability of sales, financing and capital-raising strategies, receivables policy, credit policy, debt policy, inventory policy, investment policy, tax policy, and dividend policy. Decisions in each of these spheres are monitored with the use of financial analysis tools to control the compliance of the economic processes with the adopted assumptions and to indicate how the property and financial situation can be assessed by creditors, owners, and other contractors or cooperators. The financial decisions of enterprises and their consequences remain in the sphere of interest of banks that examine the creditworthiness of enterprises and the risk factors influencing this ability in connection with financing the enterprise sector. This financing also applies to the processes of adapting enterprises to sustainable development, i.e., financing adaptation toward sustainable business models.

The sustainable business model provides benefits, but the need to develop it results from legal regulations. Other important factors influencing the sustainable business model are competition and changes in consumption patterns resulting from inter alia green consumerism. In the context of companies' financing attention should be paid to the specificity and changes in risk. This refers to the risk associated with the operation of companies, as well as the risk associated with innovative financial products creating a different type of risk, which should be considered in the financial decisions. Financing available outside the banking sector creates a risk resulting from the lack of financial supervision over selected entities, representing non-banking institutions, and offering services, e.g., lending and credit, or resulting from the distribution channels of financial services and the environment in which they are distributed.

The very risk accompanying the activities of companies may be reputation, environmental, or social risk, i.e., the risk categories that have not historically been considered by companies and which companies must learn to manage. The ESG risk present in the activities of banks and enterprises affects business models and financial situation; hence, it is crucial to manage and adapt business models to this risk. Both banks and enterprises are under the influence of regulators in terms of adjusting their activities to the requirements of sustainable development. The common fields of adjustment in this regard are office models and assets. Assets owned by enterprises are derivative of their activities and may simultaneously become a transaction security for the bank. Both the entrepreneur and the bank are interested in ensuring that the assets do not lose their value and that their valuation remains as high as possible. For enterprises, it is important from the point of view of their valuation; for the bank's activities, it is important from the point of view of transaction risk and profitability.

The strand of orphan assets appears in this context, as do brown assets. Therefore, companies that have this type of asset will become an unattractive business partner and will be forced to implement adjustment activities toward sustainability. Enterprises that do not adapt will have to consider their exclusion from the availability of financial products and services or worse terms of cooperation with banks. This primarily concerns the amount of the credit and the interest rate. Apart from assets that fit into the physical risk component, reputational risk is also significant. Both banks and entrepreneurs want to be perceived as reliable business partners and eliminate factors that may negatively impact their image. In this approach, both banks and enterprises care about the policy toward suppliers and buyers, creating sustainability friendly standards in this area. This may involve, for example, defining the condition of cooperation to only socially responsible companies (CSR) or selected through green public procurement. In this approach, the key decisions made by companies concern the analysis of suppliers, their credibility, and the course of the cooperation so far to define the terms of the trade credit granted to them and the broadly understood receivables policy, as well as the choice of liquidity enhancement instruments such as factoring, netting, cash pooling, and others. Therefore, important decisions concern not only the advances from suppliers, but also the analysis and monitoring of the dynamics and structure of receivables and the methods of their recovery.

Another element is an employee-friendly human resources policy that reduces social risk, implemented by banks and enterprises, including observance of human rights, and decent wages and working conditions. Sustainable products and services also remain a common field for the operation of banks and enterprises. Enterprises transforming their products and services into sustainable products and services need financing for this process, which they usually obtain from a bank, expecting sustainable financing from the bank, usually in the form of green loans and credits. In this case, there is a visible bilateral impact, i.e., enterprises as bank clients on the bank in the form of waiting for the bank to offer sustainable financial products and services, and banks for enterprises in the form of waiting for enterprises to implement the CSR concept or a derivative that determines a change in the business model. The common areas of influence of banks and enterprises affecting their business models are shown in Table 8.2.

In Table 8.2, the commonality is building relationships with customers, which is related to the actions of banks toward enterprises. These relationships are diagnosed and built through customer segmentation. In the context of sustainability, it is a separate customer segment. Enterprises take similar actions toward their clients, which consist of implementing CSR strategies. The activities of enterprises in building relationships with customers and segmentation are important for the bank in terms of the impact of these activities and decisions on the financial situation of the enterprise.

As part of the cooperation with clients, financial institutions are able to define the terms of this cooperation (financial and non-financial), considering the principles of partnership, ethics, and good business practices, and under the conditions dictated by the competitive market. The terms of cooperation depend on many factors, including on the type of client,

Table 8.2 The common scopes of influence of banks and enterprises affecting their business models

The scope of influence	*Bank*	*Company*	*Element of sustainable business model*
Assets (green, brown, orphan)	x	x	Key resources
Sustainable products and services	x	x	Key activities
Staff policy	x	x	Key resources
Supplier policy	x	x	Key partners
Financial condition			Cost structure/Revenue stream
ESG risk	x	x	Key resources
Distribution channels of products and services	x	x	Channels
Customer relationship	x	x	Customer relationship
Customer segments	x	x	Customer segments

Source Own elaboration

scope of cooperation, risk, and profitability. When determining transaction risk assessment criteria, financial institutions select entities, excluding from the availability of financial services those entities that do not meet the requirements they specify. However, entities meeting the criteria, depending on the assessment of the transaction risk level and conditioned by the degree of fulfillment of the criterion, are differentiated by financial institutions in terms of the conditions of using the services. Therefore, as financial institutions decide on the availability of capital to enterprises, they have the ability to influence the behavior of entrepreneurs, including motivating them to transform business models toward sustainability.

8.3 EU Taxonomy for Sustainable Activities Versus Companies' Sustainable Business Models

The taxonomy created by the European Union (the EU taxonomy) is intended to support the implementation of green governance while classifying companies according to their impact on adapting to or mitigating climate change. The scope of the taxonomy establishes technical criteria for designated economic activities. The adoption of the criteria and their fulfilment are intended to mitigate climate change while adapting organizations to climate change. The taxonomy notes that the economic activities included in the taxonomy must adopt at least one of the indicated goals, while supporting the others and not harming them (meeting minimum safety standards). The taxonomy contains very restrictive technical criteria. The European Union sees the need to invest in sustainable projects, among other, because of the need to protect not only the environment but also societies from the negative consequences of climate change, as well as other threats. An important element is the creation of actions within the entire EU, as well as individual countries, that will support sustainable growth. The guidelines and plans in this area also refer to the methods of financing such actions. To support these initiatives, a common system for classifying sustainable economic activities was created, called the EU taxonomy (https://ec.europa.eu/info/business-economy-euro/banking-and-finance/sustainable-finance/eu-taxonomy-sustainable-activities_en). The taxonomy is a set of new regulations related to sustainability and finance. Its implementation is aimed at directing the flow of capital into sectors of the economy that are conducive to sustainable development. It will affect both the financial market and the real sphere of economy, including the corporate sector.

The taxonomy (the following excerpt is based on https://ec.europa.eu/info/business-economy-euro/banking-and-finance/sustainable-finance/eu-taxonomy-sustainable-activities_en, Sustainable finance taxonomy—Regulation (EU) 2020/852. Taxonomy: Final report of the Technical Expert, Group on Sustainable Finance, March 2020 and Taxonomy report: technical annex, Updated methodology and Updated Technical Screening Criteria, March 2020) is a classification system that provides a list of identified and defined environmentally sustainable business activities. This document is intended to serve and support the implementation of the European Green Deal and increase the amount of green and sustainable investments. The developed taxonomy contains a set of definitions, based on which an activity can be easily and unambiguously classified as environmentally sustainable. The document is intended to be an aid in implementing green practices, assisting organizations in transitioning to more sustainable management strategies, and identifying sectors that require the most support in this regard. The taxonomy is governed by a legal regulation that includes commitments to identify actual environmentally sustainable activities by defining technical screening criteria for each environmental objective through delegated acts.

The taxonomy establishes six specific environmental objectives:

- climate change mitigation;
- climate change adaptation;
- the sustainable use and protection of water and marine resources;
- the transition to a circular economy;
- pollution prevention and control;
- the protection and restoration of biodiversity and ecosystems.

The purpose of the taxonomy is to help plan and report the transition to a more sustainable economy in line with the EU's environmental objectives. The reporting obligations it identifies are intended and designed to report on progress toward meeting specific selected criteria and on the achievement of those criteria. The taxonomy defines tools for financing the transformation of economies toward clearly defined environmental goals. The indicated criteria assist investors, entrepreneurs, and the financial sector in understanding the necessary pace and depth of change and the necessary transformations in implementing pro-environmental solutions. The taxonomy is in line with the European Green Deal, which has created an overarching framework and action programs to support the transformation of the European economy to a more sustainable one. Notably, the main goal is to be climate-neutral by 2050. The taxonomy's objectives and technical indicators provide an important contribution to climate change mitigation and adaptation, while meeting other sustainability objectives, namely, sustainable use and protection of water and marine resources, transition to a circular economy, pollution prevention control, and protection and restoration of biodiversity and ecosystems (Table 8.3). The taxonomy is divided into two parts: sectors that provide a substantial contribution to climate change mitigation, which includes an indication of whether the technical screening criteria relate to the performance of the economic activity or whether the activity is an enabling activity; and activities that substantially contribute to climate change adaptation.

The Table 8.3 may and will be subject to certain modifications. The presented version is valid, but at the beginning of 2021 the Sustainable Finance Platform recommended additions and possible changes to the table, which are to be introduced gradually over the time horizons of 2025 and 2050. The proposals in the document [Platform on Sustainable Finance, Consultation Response, Taxonomy draft Delegated Act setting out technical screening criteria for climate change mitigation and adaptation] refer, among other things, to updating NACE to ensure alignment

Table 8.3 EU taxonomy

Classification		*Environmental contributions*							
NACE macro-sector	*NACE activity*	*1. Climate change mitigation (Substantial Contribution)*			*2. Climate change adaptation (DNSH)*	*3. Water (DNSH)*	*4. Circular economy (DNSH)*	*5. Pollution (DNSH*	*6. Ecosystems (DNSH)*
		Own performance	*Enabling activities*	*Transitional activities*					
Forestry	Afforestation	✓			✓	✓		✓	✓
	Rehabilitation, Reforestation	✓			✓	✓		✓	✓
	Reforestation	✓			✓	✓		✓	✓
	Existing forest management	✓			✓	✓		✓	✓
	Conservation forest	✓			✓	✓		✓	✓
Agriculture	Growing of perennial crops	✓		✓	✓	✓	✓	✓	✓
	Growing of non-perennial crops	✓		✓	✓	✓	✓	✓	✓
	Livestock production	✓		✓	✓	✓	✓	✓	✓
Manufacturing	Manufacture of low carbon technologies		✓		✓	✓	✓	✓	✓
	Manufacture of Cement	✓		✓	✓	✓	✓	✓	✓
	Manufacture of Aluminum	✓		✓	✓	✓	✓	✓	✓

(continued)

Table 8.3 (continued)

Classification		*Environmental contributions*							
NACE macro-sector	*NACE activity*	*1. Climate change mitigation (Substantial Contribution)*			*2. Climate change adaptation (DNSH)*	*3. Water (DNSH)*	*4. Circular economy (DNSH)*	*5. Pollution (DNSH*	*6. Ecosystems (DNSH)*
		Own performance	*Enabling activities*	*Transitional activities*					
	Manufacture of Iron and Steel	✓		✓	✓	✓	✓	✓	✓
	Manufacture of Hydrogen	✓			✓	✓	✓	✓	✓
	Manufacture of other inorganic basic chemicals—Manufacture of carbon black	✓		✓	✓	✓	✓	✓	✓
	Manufacture of other inorganic basic chemicals—Manufacture of disodium carbonate (soda ash)	✓		✓	✓	✓	✓	✓	✓

Classification		*Environmental contributions*							
NACE macro-sector	*NACE activity*	*1. Climate change mitigation (Substantial Contribution)*			*2. Climate change adaptation (DNSH)*	*3. Water (DNSH)*	*4. Circular economy (DNSH)*	*5. Pollution (DNSH*	*6. Ecosystems (DNSH)*
		Own performance	*Enabling activities*	*Transitional activities*					
	Manufacture of other inorganic basic chemicals—Manufacture of chlorine	✓		✓	✓	✓	✓	✓	✓
	Manufacture of other organic basic chemicals	✓		✓	✓	✓	✓	✓	✓
	Manufacture of fertilizers and nitrogen compounds	✓		✓	✓	✓	✓	✓	✓
	Manufacture of plastics in primary form	✓		✓	✓	✓	✓	✓	✓

(continued)

Table 8.3 (continued)

Classification		*Environmental contributions*							
NACE macro-sector	*NACE activity*	*1. Climate change mitigation (Substantial Contribution)*			*2. Climate change adaptation (DNSH)*	*3. Water (DNSH)*	*4. Circular economy (DNSH)*	*5. Pollution (DNSH*	*6. Ecosystems (DNSH)*
		Own performance	*Enabling activities*	*Transitional activities*					
Electricity, gas, stream and air conditioning supply	Production of Electricity from Solar PV	✓			✓		✓		✓
	Production of Electricity from Concentrated Solar Power	✓			✓	✓	✓		✓
	Production of Electricity from Wind Power	✓			✓	✓	✓		✓
	Production of Electricity from Ocean Energy	✓			✓		✓	✓	✓

Classification		*Environmental contributions*							
NACE macro-sector	*NACE activity*	*1. Climate change mitigation (Substantial Contribution)*			*2. Climate change adaptation (DNSH)*	*3. Water (DNSH)*	*4. Circular economy (DNSH)*	*5. Pollution (DNSH*	*6. Ecosystems (DNSH)*
		Own performance	*Enabling activities*	*Transitional activities*					
	Production of Electricity from Hydropower	✓		✓	✓	✓		✓	✓
	Production of Electricity from Geothermal	✓		✓	✓	✓		✓	✓
	Production of Electricity from Gas (not exclusive to natural gas)	✓		✓	✓	✓		✓	✓
	Production of Electricity from Bioenergy (Biomass, Biogas and Biofuels)	✓		✓	✓	✓	✓	✓	✓

(continued)

Table 8.3 (continued)

Classification		*Environmental contributions*							
NACE macro-sector	*NACE activity*	*1. Climate change mitigation (Substantial Contribution)*			*2. Climate change adaptation (DNSH)*	*3. Water (DNSH)*	*4. Circular economy (DNSH)*	*5. Pollution (DNSH*	*6. Ecosystems (DNSH)*
		Own performance	*Enabling activities*	*Transitional activities*					
	Transmission and Distribution of Electricity	✓	✓		✓	✓	✓	✓	✓
	Storage of Electricity		✓		✓		✓		✓
	Storage of Thermal Energy		✓		✓		✓		✓
	Storage of Hydrogen		✓		✓		✓		✓
	Manufacture of Biogas or Biofuels	✓		✓	✓	✓	✓	✓	✓
	Retrofit of Gas Transmission and Distribution Networks	✓			✓	✓	✓	✓	✓
	District Heating/ Cooling Distribution	✓			✓	✓		✓	✓

Classification		*Environmental contributions*								
NACE macro-sector	*NACE activity*	*1. Climate change mitigation (Substantial Contribution)*			*2. Climate change adaptation (DNSH)*	*3. Water (DNSH)*	*4. Circular economy (DNSH)*	*5. Pollution (DNSH*	*6. Ecosystems (DNSH)*
		Own performance	*Enabling activities*	*Transitional activities*					
	Installation and operation of Electric Heat Pumps	✓			✓	✓			
	Cogeneration of Heat/cool and Power from Concentrated Solar Power	✓			✓	✓	✓		✓
	Cogeneration of Heat/Cool and Power from Geo—thermal Energy	✓		✓	✓	✓		✓	✓
	Cogeneration of Heat/Cool and Power from Gas (not exclusive to natural gas)	✓		✓	✓	✓		✓	✓

(continued)

Table 8.3 (continued)

Classification		*Environmental contributions*							
NACE macro-sector	*NACE activity*	*1. Climate change mitigation (Substantial Contribution)*			*2. Climate change adaptation (DNSH)*	*3. Water (DNSH)*	*4. Circular economy (DNSH)*	*5. Pollution (DNSH*	*6. Ecosystems (DNSH)*
		Own performance	*Enabling activities*	*Transitional activities*					
	Cogeneration of Heat/Cool and Power from Bioenergy (Biomass, Biogas, Biofuels)	✓		✓	✓	✓	✓	✓	✓
	Production of Heat/cool from Concentrated Solar Power	✓			✓	✓	✓		✓
	Production of Heat/cool from Geothermal	✓		✓	✓	✓		✓	✓
	Production of Heat/Cool from Gas (not exclusive to natural gas)	✓		✓	✓	✓		✓	✓

Classification		*Environmental contributions*								
NACE macro-sector	*NACE activity*	1. *Climate change mitigation (Substantial Contribution)*			2. *Climate change adaptation (DNSH)*	3. *Water (DNSH)*	4. *Circular economy (DNSH)*	5. *Pollution (DNSH*	6. *Ecosystems (DNSH)*
		Own performance	*Enabling activities*	*Transitional activities*					
	Production of Heat/cool from Bioenergy (Biomass, Biogas, Biofuels)	✓		✓	✓	✓	✓	✓	✓
	Production of Heat/cool using Waste Heat	✓			✓		✓	✓	✓
Water, sewerage, waste and remediation	Water collection, treatment and supply	✓			✓	✓			✓
	Centralized wastewater treatment	✓			✓			✓	✓
	Anaerobic Digestion of Sewage sludge	✓			✓			✓	

(continued)

Table 8.3 (continued)

Classification		*Environmental contributions*							
NACE macro-sector	*NACE activity*	*1. Climate change mitigation (Substantial Contribution)*			*2. Climate change adaptation (DNSH)*	*3. Water (DNSH)*	*4. Circular economy (DNSH)*	*5. Pollution (DNSH*	*6. Ecosystems (DNSH)*
		Own performance	*Enabling activities*	*Transitional activities*					
	Separate collection and transport of non -hazardous waste in source segregated fractions	✓			✓		✓	✓	
	Anaerobic digestion of bio-waste	✓			✓			✓	
	Composting of bio-waste	✓			✓			✓	
	Material recovery from non-hazardous waste	✓			✓				
	Landfill gas capture and utilization	✓			✓			✓	
	Direct Air Capture of CO_2		✓		✓	✓		✓	✓

Classification		*Environmental contributions*							
NACE macro-sector	*NACE activity*	*1. Climate change mitigation (Substantial Contribution)*			*2. Climate change adaptation (DNSH)*	*3. Water (DNSH)*	*4. Circular economy (DNSH)*	*5. Pollution (DNSH*	*6. Ecosystems (DNSH)*
		Own performance	*Enabling activities*	*Transitional activities*					
	Capture of anthropogenic emissions		✓		✓	✓	✓	✓	✓
	Transport of CO_2		✓		✓	✓		✓	✓
	Permanent Sequestration of captured CO_2	✓			✓	✓		✓	✓
Transportation and storage	Passenger Rail Transport (Interurban)	✓		✓	✓		✓	✓	
	Freight Rail Transport	✓		✓	✓		✓	✓	
	Public transport	✓		✓	✓		✓	✓	
	Infrastructure for low carbon transport (land transport)		✓		✓	✓	✓	✓	✓

(continued)

Table 8.3 (continued)

Classification		*Environmental contributions*							
NACE macro-sector	*NACE activity*	*1. Climate change mitigation (Substantial Contribution)*			*2. Climate change adaptation (DNSH)*	*3. Water (DNSH)*	*4. Circular economy (DNSH)*	*5. Pollution (DNSH*	*6. Ecosystems (DNSH)*
		Own performance	*Enabling activities*	*Transitional activities*					
	Passenger cars and commercial vehicles	✓		✓	✓		✓	✓	
	Freight transport services by road	✓		✓	✓		✓	✓	
	Interurban scheduled road transport	✓		✓	✓		✓	✓	
	Inland passenger water transport	✓		✓	✓	✓	✓	✓	✓
	Inland freight water transport	✓		✓	✓	✓	✓	✓	✓
	Infrastructure for low carbon transport (water transport)		✓		✓	✓	✓	✓	✓
Construction and real estate activities	Construction and real estate activities	✓			✓	✓	✓	✓	✓

Classification		*Environmental contributions*							
NACE macro-sector	*NACE activity*	1. *Climate change mitigation (Substantial Contribution)*			2. *Climate change adaptation (DNSH)*	3. *Water (DNSH)*	4. *Circular economy (DNSH)*	5. *Pollution (DNSH*	6. *Ecosystems (DNSH)*
		Own performance	*Enabling activities*	*Transitional activities*					
	Building renovation	✓		✓	✓	✓	✓	✓	✓
	Individual renovation measures, installation of renewables on -site and professional, scientific and technical activities	✓			✓			✓	
	Acquisition and ownership of buildings	✓		✓	✓			✓	✓
Information and communications	Data processing, hosting and related activities	✓		✓	✓		✓	✓	

(continued)

Table 8.3 (continued)

Classification		*Environmental contributions*							
NACE macro-sector	*NACE activity*	*1. Climate change mitigation (Substantial Contribution)*			*2. Climate change adaptation (DNSH)*	*3. Water (DNSH)*	*4. Circular economy (DNSH)*	*5. Pollution (DNSH*	*6. Ecosystems (DNSH)*
		Own performance	*Enabling activities*	*Transitional activities*					
	Data-driven climate change monitoring solutions		✓		✓				

Technical screening criteria: substantial contribution to climate change adaptation

NACE macro-sector	*NACE activity*	*1. Climate change mitigation (Substantial Contribution)*	*2. Climate change adaptation (DNSH)*	*3. Water (DNSH)*	*4. Circular economy (DNSH)*	*5. Pollution (DNSH*	*6. Ecosystems (DNSH)*
Forestry	Afforestation	✓	✓	✓		✓	✓
	Rehabilitation, Reforestation	✓	✓	✓		✓	✓
	Reforestation	✓	✓	✓		✓	✓

Technical screening criteria: substantial contribution to climate change adaptation

NACE macro-sector	*NACE activity*	*1. Climate change mitigation (Substantial Contribution)*	*2. Climate change adaptation (DNSH)*	*3. Water (DNSH)*	*4. Circular economy (DNSH)*	*5. Pollution (DNSH*	*6. Ecosystems (DNSH)*
	Existing forest management	✓	✓	✓		✓	✓
	Conservation forest	✓	✓	✓		✓	✓
Agriculture	Growing of perennial crops	✓	✓	✓	✓	✓	✓
	Growing of non-perennial crops	✓	✓	✓	✓	✓	✓
	Livestock production	✓	✓	✓	✓	✓	✓
Manufacturing	Manufacture of low carbon technologies	✓	✓	✓	✓	✓	✓
	Manufacture of Cement	✓	✓	✓	✓	✓	✓
	Manufacture of Aluminum	✓	✓	✓	✓	✓	✓
	Manufacture of Iron and Steel	✓	✓	✓	✓	✓	✓

(continued)

Table 8.3 (continued)

Technical screening criteria: substantial contribution to climate change adaptation

NACE macro-sector	*NACE activity*	*1. Climate change mitigation (Substantial Contribution)*	*2. Climate change adaptation (DNSH)*	*3. Water (DNSH)*	*4. Circular economy (DNSH)*	*5. Pollution (DNSH*	*6. Ecosystems (DNSH)*
	Manufacture of Hydrogen	✓	✓	✓	✓	✓	✓
	Manufacture of other inorganic basic chemicals—Manufacture of carbon black	✓	✓	✓	✓	✓	✓
	Manufacture of other inorganic basic chemicals—Manufacture of disodium carbonate (soda ash)	✓	✓	✓	✓	✓	✓
	Manufacture of other inorganic basic chemicals—Manufacture of chlorine	✓	✓	✓	✓	✓	✓

Technical screening criteria: substantial contribution to climate change adaptation

NACE macro-sector	*NACE activity*	1. *Climate change mitigation (Substantial Contribution)*	2. *Climate change adaptation (DNSH)*	3. *Water (DNSH)*	4. *Circular economy (DNSH)*	5. *Pollution (DNSH*	6. *Ecosystems (DNSH)*
	Manufacture of other organic basic chemicals	✓	✓	✓	✓	✓	✓
	Manufacture of fertilizers and nitrogen compounds	✓	✓	✓	✓	✓	✓
	Manufacture of plastics in primary form	✓	✓	✓		✓	✓
Electricity, gas, stream and air conditioning supply	Production of Electricity from Solar PV	✓	✓	✓		✓	✓
	Production of Electricity from Concentrated Solar Power	✓	✓	✓		✓	✓

(continued)

Table 8.3 (continued)

Technical screening criteria: substantial contribution to climate change adaptation							
NACE macro-sector	*NACE activity*	*1. Climate change mitigation (Substantial Contribution)*	*2. Climate change adaptation (DNSH)*	*3. Water (DNSH)*	*4. Circular economy (DNSH)*	*5. Pollution (DNSH*	*6. Ecosystems (DNSH)*
	Production of Electricity from Wind Power	✓	✓	✓		✓	✓
	Production of Electricity from Ocean Energy	✓	✓	✓		✓	✓
	Production of Electricity from Hydropower	✓	✓	✓	✓	✓	✓
	Production of Electricity from Geothermal	✓	✓	✓	✓	✓	✓
	Production of Electricity from Gas (not exclusive to natural gas)	✓	✓	✓	✓	✓	✓

Technical screening criteria: substantial contribution to climate change adaptation							
NACE macro-sector	*NACE activity*	*1. Climate change mitigation (Substantial Contribution)*	*2. Climate change adaptation (DNSH)*	*3. Water (DNSH)*	*4. Circular economy (DNSH)*	*5. Pollution (DNSH*	*6. Ecosystems (DNSH)*
	Production of Electricity from Bioenergy (Biomass, Biogas and Biofuels)	✓	✓	✓	✓	✓	✓
	Transmission and Distribution of Electricity	✓	✓	✓	✓	✓	✓
	Storage of Electricity	✓	✓	✓	✓	✓	✓
	Storage of Thermal Energy	✓	✓	✓	✓	✓	✓
	Storage of Thermal Energy	✓	✓	✓	✓	✓	✓
	Storage of Hydrogen	✓	✓	✓	✓	✓	✓
	Manufacture of Biogas or Biofuels	✓	✓	✓	✓	✓	✓

(continued)

Table 8.3 (continued)

Technical screening criteria: substantial contribution to climate change adaptation

NACE macro-sector	*NACE activity*	*1. Climate change mitigation (Substantial Contribution)*	*2. Climate change adaptation (DNSH)*	*3. Water (DNSH)*	*4. Circular economy (DNSH)*	*5. Pollution (DNSH*	*6. Ecosystems (DNSH)*
	Retrofit of Gas Transmission and Distribution Networks	✓	✓	✓	✓	✓	✓
	District Heating/Cooling Distribution	✓	✓	✓	✓	✓	✓
	Installation and operation of Electric Heat Pumps	✓	✓	✓	✓	✓	✓
	Cogeneration of Heat/cool and Power from Concentrated Solar Power		✓	✓	✓		✓

Technical screening criteria: substantial contribution to climate change adaptation

NACE macro-sector	*NACE activity*	*1. Climate change mitigation (Substantial Contribution)*	*2. Climate change adaptation (DNSH)*	*3. Water (DNSH)*	*4. Circular economy (DNSH)*	*5. Pollution (DNSH*	*6. Ecosystems (DNSH)*
	Cogeneration of Heat/Cool and Power from Geo—thermal Energy	✓	✓	✓		✓	✓
	Cogeneration of Heat/Cool and Power from Gas (not exclusive to natural gas)	✓	✓	✓	✓	✓	✓
	Cogeneration of Heat/Cool and Power from Bioenergy (Biomass, Biogas, Biofuels)	✓	✓	✓	✓	✓	✓

(continued)

Table 8.3 (continued)

Technical screening criteria: substantial contribution to climate change adaptation							
NACE macro-sector	*NACE activity*	*1. Climate change mitigation (Substantial Contribution)*	*2. Climate change adaptation (DNSH)*	*3. Water (DNSH)*	*4. Circular economy (DNSH)*	*5. Pollution (DNSH*	*6. Ecosystems (DNSH)*
	Production of Heat/cool from Concentrated Solar Power		✓	✓	✓		✓
	Production of Heat/cool from Geothermal	✓	✓	✓		✓	✓
	Production of Heat/Cool from Gas (not exclusive to natural gas)	✓	✓	✓		✓	✓
	Production of Heat/cool from Bioenergy (Biomass, Biogas, Biofuels)	✓	✓	✓	✓	✓	✓

Technical screening criteria: substantial contribution to climate change adaptation

NACE macro-sector	*NACE activity*	*1. Climate change mitigation (Substantial Contribution)*	*2. Climate change adaptation (DNSH)*	*3. Water (DNSH)*	*4. Circular economy (DNSH)*	*5. Pollution (DNSH*	*6. Ecosystems (DNSH)*
	Production of Heat/cool using Waste Heat		✓		✓	✓	✓
Water, sewerage, waste and remediation	Water collection, treatment and supply		✓	✓			✓
	Centralized wastewater treatment		✓			✓	✓
	Anaerobic Digestion of Sewage sludge	✓	✓			✓	
	Separate collection and transport of non-hazardous waste in source segregated fractions		✓		✓	✓	

(continued)

Table 8.3 (continued)

Technical screening criteria: substantial contribution to climate change adaptation							
NACE macro-sector	*NACE activity*	*1. Climate change mitigation (Substantial Contribution)*	*2. Climate change adaptation (DNSH)*	*3. Water (DNSH)*	*4. Circular economy (DNSH)*	*5. Pollution (DNSH*	*6. Ecosystems (DNSH)*
	Anaerobic digestion of bio-waste	✓	✓			✓	
	Composting of bio-waste		✓			✓	
	Material recovery from non-hazardous waste	✓	✓				
	Landfill gas capture and utilization	✓	✓			✓	
	Direct Air Capture of CO_2		✓	✓		✓	✓
	Capture of anthropogenic emissions	✓	✓	✓	✓	✓	✓
	Transport of CO_2	✓	✓	✓		✓	✓

Technical screening criteria: substantial contribution to climate change adaptation

NACE macro-sector	*NACE activity*	*1. Climate change mitigation (Substantial Contribution)*	*2. Climate change adaptation (DNSH)*	*3. Water (DNSH)*	*4. Circular economy (DNSH)*	*5. Pollution (DNSH*	*6. Ecosystems (DNSH)*
	Permanent Sequestration of captured CO_2	✓	✓	✓		✓	✓
Transportation and storage	Passenger Rail Transport (Interurban)	✓	✓		✓	✓	
	Freight Rail Transport	✓	✓		✓	✓	
	Public transport	✓	✓		✓	✓	
	Infrastructure for low carbon transport (land transport)	✓	✓	✓	✓	✓	✓
	Passenger cars and commercial vehicles	✓	✓		✓	✓	
	Freight transport services by road	✓	✓		✓	✓	
	Interurban scheduled road transport	✓	✓		✓	✓	

(continued)

Table 8.3 (continued)

Technical screening criteria: substantial contribution to climate change adaptation							
NACE macro-sector	*NACE activity*	*1. Climate change mitigation (Substantial Contribution)*	*2. Climate change adaptation (DNSH)*	*3. Water (DNSH)*	*4. Circular economy (DNSH)*	*5. Pollution (DNSH*	*6. Ecosystems (DNSH)*
	Inland passenger water transport	✓	✓	✓	✓	✓	✓
	Inland freight water transport	✓	✓	✓	✓	✓	✓
	Infrastructure for low carbon transport (water transport)	✓	✓	✓	✓	✓	✓
Buildings	Construction of new buildings	✓	✓	✓	✓	✓	✓
	Building renovation	✓	✓	✓	✓	✓	✓
Financial and insurance activities	Non-life insurance		✓				

Technical screening criteria: substantial contribution to climate change adaptation

NACE macro-sector	*NACE activity*	1. *Climate change mitigation (Substantial Contribution)*	2. *Climate change adaptation (DNSH)*	3. *Water (DNSH)*	4. *Circular economy (DNSH)*	5. *Pollution (DNSH*	6. *Ecosystems (DNSH)*
Professional, scientific, and technical activities	Engineering activities and related technical consultancy dedicated to adaptation to climate change		✓				

Source Taxonomy: Final report of the Technical Expert, Group on Sustainable Finance, March 2020 and Taxonomy report: technical annex, Updated methodology and Updated Technical Screening Criteria, March 2020 with the taxonomy and reflect market needs, better alignment of sectors and their appropriate classification, and not to overlook any relevant activities of different organizations in the environmental context, indicating that the focus should not only be on those sectors that are not environmentally neutral due to their activities, but to expand the scope of the taxonomy so that entire economies move toward sustainability. Another recommendation is that the identification should be clear, for reporting purposes, so that it can be applied as widely as possible. The taxonomy should also be consistent and as useful as possible. This means that it should focus on specific adaptation strategies for all sectors, prioritizing, selecting, and paying more attention to climate risks and environmental impacts of a given sector. Recommendations for change in the resulting document are intended to help make the solution more useful and best identified. It is not entirely clear which of the recommendations will be implemented and in what time horizon. It is important that the indicated taxonomy constitutes a basis which will probably be modified and developed in the nearest future.

The economic sectors and activities included in the taxonomy to date have the potential to significantly contribute to climate change mitigation or adaptation. The taxonomy focuses on the activities with the most harmful environmental impacts. Not all industries and sectors are included and shown. The methodological approach of the taxonomy indicates that the Taxonomy Regulation (TR) recognizes two distinct types of substantial contribution that can be considered taxonomy-aligned:

- Economic activities that substantially contribute based on their own performance, for example, an economic activity being performed in a way that is environmentally sustainable;
- Enabling activities, which are economic activities that, by provision of their products or services, enable a substantial contribution in other activities, for example, an economic activity that manufactures a component that improves the environmental performance of another activity.

The Taxonomy Regulation sets out three groups of taxonomy users: financial market participants offering financial products in the EU, including occupational pension providers; large companies who are already required to provide a non-financial statement under the Non-Financial Reporting Directive; and the EU and Member States when setting public measures, standards, or labels for green financial products or green (corporate) bonds.

The indicated taxonomy supports decision-making by the financial sector. The financial industry will support the principles of sustainable development in its product offerings, acting in accordance with the principles of green transformation and green governance. The financial industry must responsibly support the indicated principles and elements of the taxonomy by aligning its products, initiatives, and engagement with the indicated guidelines. The financial sector plays a vital role in the implementation of the principles of sustainable development, supporting the activities of most organizations, including those whose activities and impacts are not climate-neutral. In such cases, companies that wish to obtain financing and support from financial institutions must explicitly transform their activities to receive such support and act in accordance with the guidelines of the taxonomy and the principles of sustainable development. The taxonomy is intended to both assist in the selection and application of appropriate financial tools and products, while financial institutions can use it as a tool to indicate which industries or sectors they will not finance or will finance under what conditions. Environmental awareness and responsibility can be one of the most important elements of collaboration between organizations and their links in the supply chain and the financial sector. Only joint action in accordance with specific guidelines and clear rules can contribute to the success of sustainable development. Industries with the highest emissions—mining, transport, processing, and construction—must change their business strategies toward more environmentally friendly ones. This is a requirement not only in line with economy-wide actions and directions of particular states' policies, but also a necessity for cooperation between various dependent sectors (the financial sector and enterprises). The financial sector can adapt to the trends and directions in economic development, and through its requirements, supported, e.g., by the assumptions of the indicated taxonomy, forces the corporate sector to introduce systemic changes. The actions of the financial sector in this area clearly indicate that it is impossible to finance activities that are incompatible with sustainable development (Report, 2020). At the same time, they can support customers through a system of training providing information and creating programs to support and help organizations with the difficult process of transformation. Green governance and sustainable development are challenges to be met, among others, by introducing activities that transform existing activities or by using the EU taxonomy. A large role in the success of the activities undertaken is played by the monitoring system, as well as the identification and mitigation of risks or the flow and transfer of information and data management. Other important aspects are education, data management, and the development of offerings to match global trends and developments.

The financial sector and enterprises as well as legal and organizational guidelines must be developed together. Business offers must be adapted to new requirements; the business sector must comply with these requirements and often reorganize itself to receive financial support; and guidelines, e.g., those of the European Union, must support all stakeholders in the implementation of sustainability through clearly defined criteria.

In addressing the impact of the taxonomy on organizations, as outlined in the Deloitte 2019 report (Deloitte, 2019), companies that are dominated by equity will not be affected by the introduction of the taxonomy. However, it may affect organizations' ability to raise funds in the future. Organizations that rely on credit and loans as sources of funding for their operations may see credit conditions deteriorate for companies with high exposure to ESG factors. A similar situation may occur in the case of equity and bond issues. Companies that choose to make investments that lead to climate change mitigation can expect increased investor interest. Organizations will need to disclose non-financial information based on complete and reliable data related to climate action and carbon performance. Organizations that are carbon-intensive will also have large potential for change and opportunities to raise capital for those purposes. However, this will require taking action and implementing innovative strategies for doing business. It will be necessary to use innovative technologies for which appropriate funds will have to be obtained. Organizations that do not introduce changes and taxonomy requirements will be less attractive for investors, but will also become less reliable and interesting partners in sustainable supply chains. Non-organic industries or companies, especially large ones, may experience organizational problems in implementing the necessary changes, which may be related to organizational structures, less flexibility, and problems with, for example, making data public, and collecting, processing, and using data as an element supporting the activity, but also as a factor subject to evaluation, on the basis of which financial institutions will make decisions on whether or not to finance the indicated activity.

Environmental aspects implemented in individual sectors or whole economies are a visible trend: environmental as well as social and corporate governance factors are beginning to play an increasingly important role. Transformation of the business sector toward a more environmental one is becoming increasingly visible and an integral part of new business strategies in both energy and infrastructure transformations (green buildings, ecological vehicle fleet, and the use of renewable energy).

Both the requirements of the created taxonomy and the observed trends indicate that greening must be integrated into all investment, credit, and production processes. The actions taken require high transparency and uniformity of criteria. This is associated with the need for good and efficient information flow. The process of data collection must be based on known and transparent principles, and must be standardized, but also created so that they are understandable and convince potential users to use them. Another important element is the general binding regulations that apply to all actors: companies have to comply with them, and the financial sector has to adapt its products to the regulations and new trends. Businesses using the taxonomy should seize the opening opportunity to grow their business in a low-carbon and climate- and biodiversity-friendly direction, and financial institutions have a basis for financing environmentally sustainable activities.

8.4 The Model of Relationship Banking Toward Common Sustainable Business Models of Banks and Companies

Previous research points to the importance of companies in creating green economic growth and sustainability; achieving SDGs and stability are important to banks. It also indicates the need to ensure that companies fulfill their role in sustainability innovation so they can continue to be a major contributing factor in the economy. This goal, without financing and sustainable financing, without banking products and services supporting sustainability goals, is impossible to achieve.

To help companies accomplish this, government institutions and financial institutions including banks should work to overcome the obstacles faced by companies in reaching this goal. Companies' main obstacle is access to and use of sources of finance (Petersen & Rajan, 1994; Beck & Demirgüc¸-Kunt, 2006; Bruns & Fletcher, 2008; Beck et al., 2008; Al-Hyari et al., 2012; Wonglimpiyarat, 2015; Badaj & Radi, 2017; Shihadeh et al., 2019).

The interaction and even the creation of relations between global financial markets in a bank and the economy influences sustainable development (Huang, 2011). For individual companies, sustainability translates into corporate social responsibility (Dahlsrud, 2008), and relationship banking is increasingly required to understand the needs and directions of environmental adaptation, including the need to reduce ESG risks. Therefore, the strategy for the development of responsible business should include the economic (profit), ecological (planet), and ESG risk and social (people) dimensions, considering relations with banks as financial institutions.

The literature on the subject discusses the special importance of the bank–client relational approach. When financial economists speak of relationships between banks and firms, they mean a close relationship between a company and its banker, in which a single banker has detailed and very deep knowledge about the firm's affairs, built up over years of cooperation (Guo, 2017). The basics of relationship banking were formulated in the works of Ongena and Smith (2000), Boot (2000), and Berger and Udell (1995). It is seen as a relationship between the bank and the companies, which is more than a simple, anonymous financial transaction (Ongena & Smith, 2000) that benefits both parties. From the bank's point of view, an advantage is defined as wider access to information about the enterprise, while the companies gain access to financing and financial services from the bank, even when experiencing financial difficulties. The relationship is based on two dimensions: time, which is the measure of the length of cooperation between the bank and the companies (duration), and the scope of cooperation, measured by the types of products and services used or wanted by the company (scope).

At this point, relationship banking may be of particular importance in developing a business based on sustainability, and will be of particular importance in reducing ESG risks for clients. One should look for strength in mutual relations and exchange of experience in the field of ESG risk mitigation.

The consensus is that banks engaging in relationship banking and, in particular, relationship lending (including Boot, 2000; Cull et al., 2011) also offer other services that do not require such close relationships or do not bind the parties in the long term. The indicated dimensions, i.e., time, scope, and the possibility of going beyond classic services, provide the basis for formulating a model of cooperation between the bank and the company perceived by the bank in ESG terms. Examples of cooperation areas in relationship banking include bank account services; effective management of capital surpluses, including green instruments; financing of the company's operations (current and investment, including pro-ecological); conducting cash settlements, investments, and ecological innovations; financial consulting, financial risk management; and ESG training in various aspects of sustainability and CSR; and accounting and the application of SRI standards. Another important element is linking the bank's strategy and business model, considering sustainability standards (including SDGs and CSR goals) with the company's business model and the elements of sustainability or those based on sustainability. Figure 8.2 presents a proposed bank–enterprise cooperation model as perceived by the bank under ESG conditions, which considers the above assumptions.

Adaptation to the implementation of SDGs and considering ESG risk causes a change in the banking offer. Banks are using relationship banking to redirect their clients toward sustainability and to collect broader information and build a knowledge base. These knowledge bases are the basis of relationship banking with the enterprise and constitute an important area of cooperation on which the enterprise is based when building its new business model. The company, depending on its awareness, perception of ESG risk, the degree of development, and/or stakeholders, may redefine three types of new business models. These business models are shown in Fig. 8.2, which describes their ideological assumptions. Note that based on relationship banking and new business models, new value is created for S stakeholders. This value is characterized by the following equation.

$$\text{New value for the customer} = \frac{\text{effects created for the client} + \text{process quality}}{\text{Price for the client} + \text{purchase cost}} + S$$

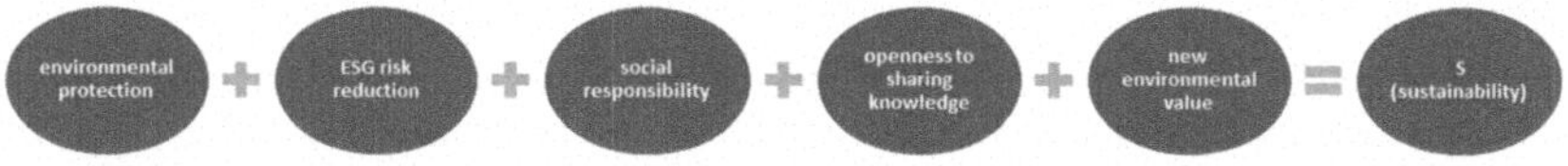

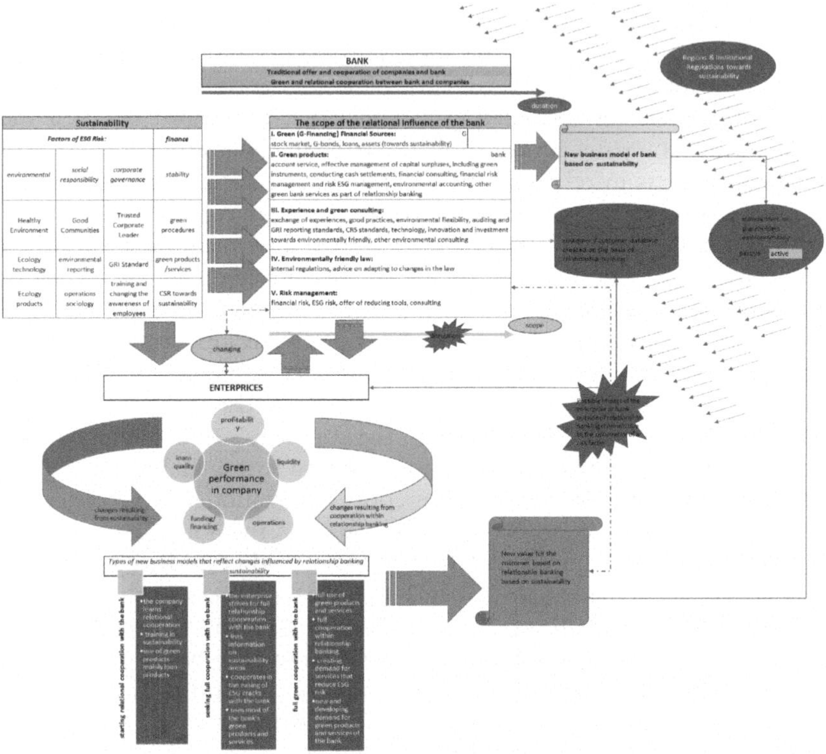

Fig. 8.2 Proposed bank–enterprise cooperation model based on green relationship banking (*Source* Own elaboration)

The new value based on relationship banking enables the development of cooperation, but also affects the banks. They change under the influence of cooperation with companies, adjusting their business models toward sustainability but are more open to cooperation with the company with which they are already in a relationship. Due to the exchange of information, they build their knowledge bases, which allow them to react to the emerging risks and develop procedures related to the mitigation of ESG risk.

The model of cooperation between the business sector and financial institutions under ESG conditions should increase the coherence of actions and synergies in already implemented initiatives, support new ideas, and be embedded in innovative business strategies. The cooperation model should allow for the benefits of collaboration, especially in the context of strengthening cooperative ties and building cross-sector partnerships. This cooperation may support the development and implementation of CSR principles and innovative business strategies, with the active support of the financial sector, as a link actively participating in such activities. Supporting new initiatives is impossible without financial resources, and flexible financial mechanisms should be encouraged and used. Such mechanisms may include the principles of public–private partnerships or mobilization of the investment community, both institutional and individual, to include environmental, social, and governance (ESG) risks in investment decisions. The activities of the corporate sector must focus on constructing strategies so that investments and all kinds of activities are responsible, support sustainable consumption (this is also the role of the sector in creating new products and services), supporting CSR, as well as increasing the transparency and credibility of initiatives. These conditions can be met by implementing modern CSR standards that are based on risk management and disclosure of ESG data. Such activities are supported by, e.g., the indicated EU taxonomy, as well as appropriate screening and reporting of the undertaken activities. To justify and prove pro-environmental and sustainable actions, ESG decisions have to be considered in investment processes and when implementing new business strategies. They are becoming one of the factors in making decisions on financial support by financial institutions; therefore, they are starting to play an increasingly important role.

The model of cooperation must also include activities that disseminate the concept of responsible investment and the principles of ESG-aware support for the business sector by financial institutions. The strategy for managing the organization, its operations, and future investments must consider both CSR aspects and the inclusion of environmental, social, and economic risks. An important element is the process of education, increasing awareness of all stakeholders in this area, as well as providing reliable and credible data, because only in this way can the requirements be met and sustainable development strategies be implemented in a clear and competent manner. Two important elements of building cooperation between all stakeholders are education and data flow. Information on the application of certain procedures, principles, or the implementation of CSR should not only be public, but should also be verified when establishing cooperation in supply chains and between individual organizations and companies. Such joint, synergic actions also provide more opportunities for the financial sector. Companies become more credible not only as a result of their actions, but also by indicating the principles on which they cooperate with other entities. This may be a certain determinant and benefit considered by the financial sector. When building the model, it is worth creating and using the recommendations and guidelines that have been developed (e.g., standards, benchmarks, and/or EU regulations). Activities in the area of modeling cooperation should be systemic to achieve the greatest possible synergy effects. The system of monitoring, evaluation, and periodic adjustments is important. These activities should also be systemic and concern all stakeholders and participating parties.

Considering ESG in all areas of activity and cooperation has an increasingly strong influence on shaping the conditions for development and the future condition of the environment and economies, especially taking into account the perspective of ensuring environmental order for future generations. The model of cooperation should also consider the promotion of good practices in terms of cooperation of the business sector with the administration and the financial sector as well as with scientific and research centers for the implementation of research responding to the real needs of economic development. Another important element is greater participation and trust, which is possible to achieve with greater involvement of all parties in the process of implementing innovative strategies, financing green investments, and acting in accordance with the requirements and regulatory legislators. An important aspect of the cooperation model is the development of platforms for cooperation, consultation, and strengthening new forms of communication.

Innovative business strategies can support the formation and positioning of the companies concerned in the international and national arenas. In addition to offering good products and services, issues of capital security, stability of organizational structures, and their reconfiguration to meet ESG requirements are becoming increasingly important. These factors will determine competitive advantages, and they can be guaranteed by companies that comply with international standards of reliability or minimize ESG risk in their entire supply chains. A financing system that fosters innovative, socially responsible, and pro-environmental solutions will allow even the most bold and innovative strategies of the business sector to be implemented. With support from the financial sector, it will be easier to implement reconfigurations, restructurings, and transitions to greener operations. The common model must include simplification and clarity of procedures. The rules of operation or financing must be consistent, transparent, as well as stable, so that they can be easily referred to and applied. The interest of financial institutions in promoting the concept of CSR, supported by the EU taxonomy, is a signal to the business sector of the direction in which changes must be made. The need for change that is promoted not only from the legislative side and regulators, but also from the financial sector may have more power and provide an incentive to undertake changes in enterprises. Lack of financing from the banking sector can lead to a lack of investment and development opportunities for businesses, so it is important to act together in a systemic way to support each of the stakeholders.

Other elements of the model are information and training activities. Their task is to provide knowledge, especially in the area of requirements and credibility assessment or ESG data analysis. The offered tools and comparative models are helpful here. Such activities allow both the financial sector to estimate the investment risk and the companies to estimate the development potential and elements that need to be modified. Socially responsible businesses must cooperate while meeting the requirements of other market participants, including financial institutions. Including social, ethical, or environmental issues in the management of organizations is becoming a mandatory element. An additional element may be the transition to more ecological production or offering of such products. Here, too, we can see the cooperation and requirements of the financial sector, and, for example, the lack of possibilities to finance activities that are not compliant with the environmental requirements.

References

Al-Hyari, K., Al-Weshah, G., & Alnsour, M. (2012). Barriers to internationalisation in SMEs: Evidence from Jordan. *Marketing Intelligence and Planning, 30*(2), 188–211. https://doi.org/10.1108/02634501211211975.

Badaj, F., & Radi, B. (2017). Empirical investigation of SMEs' perceptions towards PLS financing in Morocco. *International Journal of Islamic and Middle Eastern Finance and Management, 11*, 250–273. https://doi.org/10.1108/IMEFM-05-2017-0133.

Beck, T., & Demirgüc¸-Kunt, A. (2006). Small and medium-size enterprises: Access to finance as a growth constraint. *Journal of Banking and Finance, 30*(11), 2931–2943. https://doi.org/10.1016/j.jbankfin.2006.05.009.

Beck, T., Demirgüç-Kunt, A., & Peria, S. M. (2008). Banking services for everyone? Barriers to bank access and use around the world. *World Bank Economic Review, 22*, 397–430.

Berger, A. N., & Udell, G. F. (1995). Relationship lending and lines of credit in small firm finance. *Journal of Business, 68*, 351–381.

Boot, A. W. A. (2000). Relationship banking: What do we know. *The Journal of Financial Intermediation, 9*, 7–25.

Bruns, V., & Fletcher, M. (2008). Banks' risk assessment of Swedish SMEs. *Venture Capital, 10*(2), 171–194. https://doi.org/10.1080/13691060801946089.

Cull, R., Haber, S., & Imai, M. (2011). *Related lending and banking development, policy* (Research Working Paper, No. 5570). The World Bank Development Research Group, Finance and Private Sector Development Team.

Dahlsrud, A. (2008). How corporate social responsibility is defined. *Corporate Social Responsibility and Environmental Management, 15*, 1–13.

Deloitte. (2019). *Perspektywy rozwoju zrównoważonego finansowania – implikacje dla sektora przedsiębiorstw finansowych.*

Global4. (2020). *Business model canvas – Global4net.* https://global4net.com/ecommerce/business-model-canvas-co-to-jest-i-jak-go-stworzyc/.

Guo, Y. (2017). Implementing relationship banking strategies and techniques and improving customer value. *Finance and Market, 2*(2), 1–13.

https://ec.europa.eu/info/business-economy-euro/banking-and-finance/sustainable-finance/eu-taxonomy-sustainable-activities_en.

Huang, Y. (2011). Is economic volatility detrimental to global sustainability? *World Bank Economic Review, 26*, 128–146.

Klimontowicz, M., & Pyka, A. (2016). Banks' business models – The challenges of XXI century. *Studia i Prace WNEiZ US, 44*(2), 153–165. https://doi.org/10.18276/sip.2016.44/2-12.

KPMG International. (2011, December). *Evolving banking regulation. A long journey ahead – the outlook for 2012.* Report.

Ongena, S., & Smith, D. C. (2000). Bank relationships: A survey. In P. Haker & S. A. Zenios (Eds.), *The performance of financial institutions* (pp. 221–258). Cambrige University Press.

Petersen, M. A., & Rajan, R. G. (1994). The benefits of lending relationships: Evidence from small business data. *The Journal of Finance, 49*(1), 3–37. https://doi.org/10.2307/2329133.

Raport 2020: Zielone finanse - nowe spojrzenie na finansowanie. (2020). Klub Odpowiedzialnych Finansów przy Europejskim Kongresie finansowym, Sopot.

Scholtens, B., & van't Klooster, S. (2019). Sustainability and bank risk. *Palgrave Communications, 5*, 105. https://doi.org/10.1057/s41599-019-0315-9.

Shihadeh, F., Gamage, S. K. N., & Hannoon, A. M. T.(2019). The causal relationship between SME sustainability and banks' risk. *Economic Research-Ekonomska Istraživanja, 32*(1), 2743–2760. https://doi.org/10.1080/1331677X.2019.1655465.

Sustainable finance taxonomy—Regulation (EU) 2020/852.

Taxonomy report: Technical annex, Updated methodology and Updated Technical Screening Criteria, March 2020

TEG final report on the EU taxonomy, Taxonomy: Final report of the Technical Expert, Group on Sustainable Finance, European Commission, March 2020

Wonglimpiyarat, J. (2015). Challenges of SMEs innovation and entrepreneurial financing. *World Journal of Entrepreneurship, Management and Sustainable Development, 11*(4), 295–311. https://doi.org/10.1108/WJEMSD-04-2015-0019.

References

Aagaard, A. (Ed.). (2018). *Sustainable business models: Innovation.* Implementation and Success: Springer.

Abdelkafi, N., & Täuscher, K. (2016). Business models for sustainability from a system dynamics perspective. *Organization & Environment, 29*(1), 74–96.

Abdelkafi, N., Makhotin, S., & Posselt, T. (2013). Business model innovation for electric mobility: What can be learned from existing business model patterns? *International Journal of Innovation Management, 17*, 1–41. https://doi.org/10.1142/S1363919613400033.

Abraham, S. (2013). Will business model innovation replace strategic analysis? *Strategy & Leadership, 41*(2), 31–38.

Achtenhagen, L., Melin, L., & Naldi, L. (2013). Dynamics of business models–strategizing, critical capabilities and activities for sustained value creation. *Long Range Planning, 46*(6), 427–442.

Afuah, A. (2004). *Business models: A strategic management approach.* Boston, MA: McGraw-Hill, Irwin.

Agenda for Sustainable Development. (2019). https://sustainabledevelopment.un.org/sdgsummit. Accessed 31 Dec 2020.

Agudo-Valiente, J. M., Garcés-Ayerbe, C., & Salvador-Figueras, M. (2017). Corporate social responsibility drivers and barriers according to managers' perception; Evidence from Spanish firms. *Sustainability, 9*(10), 1821.

Aithal, P. S. (2016). A review on various E-business and M-business models & research opportunities. *International Journal of Management, IT and Engineering, 6*(1), 275–299.

Alińska, A., Filipiak, B. Z., & Kosztowniak, A. (2018). The importance of the public sector in sustainable development in Poland. *Sustainability, 10*, 3278.

© The Editor(s) (if applicable) and The Author(s), under exclusive license to Springer Nature Switzerland AG 2021

M. Ziolo et al., *Sustainability in Bank and Corporate Business Models*, Palgrave Studies in Impact Finance,
https://doi.org/10.1007/978-3-030-72098-8

Al-Hyari, K., Al-Weshah, G., & Alnsour, M. (2012). Barriers to internationalisation in SMEs: Evidence from Jordan. *Marketing Intelligence & Planning, 30*(2), 188–211. https://doi.org/10.1108/02634501211211975.

Allen, J. C., & Malin, S. (2008). Green entrepreneurship: A method for managing natural resources? *Society & Natural Resources: An International Journal, 21*(9), 828–844.

Alsayegh, M. F., Rahman, R., & Homayoun, S. (2020). Corporate economic, environmental, and social sustainability performance transformation through ESG disclosure. *Sustainability, 12,* 3910. https://doi.org/10.3390/su12093910.

Ambec, S., & Lanoie, P. (2008). Does it pay to be green? A systematic overview. *Academy of Management Perspectives, 22,* 45–62.

Amit, R., & Zott, C. (2001). Value creation in e-business. *Strategic Management Journal, 22,* 493–520.

Amshoff, B., Dülme, C., Echterfeld, J., & Gausemeier, J. (2015). Business model patterns for disruptive technologies. *International Journal of Innovation Management, 19*(03), 1540002.

Arestis, P. (2006). Financial liberalization and the relationship between finance and growth. In P. Arestis & M. Sawyer (Eds.), *A handbook of alternative monetary economics* (pp. 346–364). Edward Elgar Publishing.

Avdiji, H., Elikan, D., Missonier, S., & Pigneur, Y. (2018). Designing tools for collectively solving III-structured problems. *HICSS, 2018,* 1–10.

Ayadi, R., Cucinelli, D., & de Groen, W. P. (2019). *The banking business models (BBM) monitor 2019 Europe: Performance, risk, response to regulation and resolution: 2005–2017* (pp. 65–101). London: Cass Business School.

Badaj, F., & Radi, B. (2017). Empirical investigation of SMEs'perceptions towards PLS financ-ing in Morocco. *International Journal of Islamic and Middle Eastern Finance and Management, 11,* 250–273. https://doi.org/10.1108/IMEFM-05-2017-0133.

Baden-Fuller, C., & Haefliger, S. (2013). Business models and technological innovation. *Long Range Planning, 46,* 419–426. https://doi.org/10.1016/j.lrp.2013.08.023.

Baden-Fuller, C., & Morgan, M. S. (2010). Business models as models. Long Range Planning, 43(2–3), 156–171. https://doi.org/10.1016/j.lrp.2010.02.005.

Baden-Fuller, C., & Mangematin, V. (2013). Business models: A challenging agenda. *Strategic Organization, 11*(4), 418–427.

Baldassarre, B., Calabretta, G., Bocken, N. M. P., & Jaskiewicz, T. (2017). Bridging sustainable business model innovation and user-driven innovation: A process for sustainable value proposition design. *Journal of Cleaner Production, 147,* 175–186.

Ballantyne, D., Frow, P., Varey, R. J., & Payne, A. (2011). Value propositions as communication practice: Taking a wider view. *Industrial Marketing Management, 40*(2), 202–210.

Ban Track. (2006). https://www.banktrack.org/news/the_dos_and_donts_of_sustainable_banking. Accessed 12 Nov 2020.

Bansal, P., & Roth, K. (2000). Why companies go green: A model of ecological responsiveness. *Academy of Management Journal, 43*(4), 717–736.

Bansal, P., & DesJardine, M. R. (2014). Business sustainability: It is about time. *Strategic Organization, 12*(1), 70–78.

Barbieri, R., & Santos, D. F. L. (2020). Sustainable business models and eco-innovation: A life cycle assessment. *Journal of Cleaner Production, 266.*

Barral, V. (2012, May). Sustainable development in international law: nature and operation of an evolutive legal norm. *European Journal of International Law, 23*(2), 377–400. https://doi.org/10.1093/ejil/chs016.

Baumann-Pauly, D., Wickert, C., Spence, L. J., & Scherer, A. G. (2013). Organizing corporate social responsibility in small and large firms: Size matters. *Journal of Business Ethics, 115*(4), 693–705.

Becchetti, L., Garcia, M., & Trovato, G. (2011). Credit rationing and credit view: Empirical evidence from loan data. *Journal of Money, Credit and Banking, 43*(6), 1217–1245.

Beck, T., & Demirgüc¸-Kunt, A. (2006). Small and medium-size enterprises: Access to financeas a growth constraint. *Journal of Banking & Finance, 30*(11), 2931–2943. https://doi.org/10.1016/j.jbankfin.2006.05.009.

Beck, T., Demirgüç-Kunt, A., & Peria, S. M. (2008). Banking services for everyone? Barriers to bank access and use around the world. *World Bank Economic Review, 22*, 397–430.

Bellman, R., Clark, Ch., Craft, C., Mlcolm, D. G., & Riccardi, F. (1957). On the construction of a multi-stage, multi-person business game. Operations Research, The Rand Corporation, California, 469–503.

Belz, F. M., & Binder, J. K. (2017). Sustainable entrepreneurship: A convergent process model. *Business Strategy and the Environment, 26*(1), 1–17.

Benedikter, R. (2011). *European answers to the financial crisis: Social banking and social finance, spice digest*. Stanford University Press.

Berger, A. N., & Udell, G. F. (1995). Relationship lending and lines of credit in small firm finance. *Journal of Business, 68*, 351–381.

Bian, J., Liao, Y., Wang, Y. Y., & Tao, F. (2020). Analysis of firm CSR strategies. *European Journal of Operational Research, 290*(3), 914–926.

Biswas, N. (2011). Sustainable green banking approach, the need of the hour. *Business Spectrum, 1*(1), 32–38.

Bocken, N. M. P., Short, S. W., Rana, P., & Evans, S. (2014). A literature and practice review to develop sustainable business model archetypes. *Journal of Cleaner Production, 65*, s.42–56.

Bocken, N. M. P., Rana, P., & Short, S. W. (2015). Value mapping for sustainable business thinking. *Journal of Industrial and Production Engineering, 32*(1), 67–81.

Bocken, N., Short, S., Rana, P., & Evans, S. (2013). A value mapping tool for sustainable business modelling. *Corporate Governance, 13,* 482–497.

Bolis, I., Morioka, S. N., & Sznelwar, L. I. (2014). When sustainable development risks losing its meaning. Delimiting the concept with a comprehensive literature review and a conceptual model. *Journal of Cleaner Production, 83,* 7–20.

Boons, F., & Lüdeke-Freund, F. (2013). Business models for sustainable innovation: State-of-the-art and steps towards a research agenda. *Journal of Cleaner Production, 45,* 9–19.

Boot, A. W. A. (2000). Relationship banking: What do we know. *The Journal of Financial Intermediation, 9,* 7–25.

Boudreau, M. C., Chen, A., & Huber, M. (2008). Green IS: Building sustainable business practices. *Information Systems: A Global Text,* 1–17. https://www.researchgate.net/publication/242112741_Green_IS_Building_Sustainable_Business_Practices. Accessed Dec 2020.

Bowman, C., & Ambrosini, V. (2000). Value creation versus value capture: Towards a coherent definition of value in strategy. *British journal of management, 11*(1), 1–15.

Bradley, P., Parry, G., & O'Regan, N. (2020). A framework to explore the functioning and sustainability of business models. *Sustainable Production and Consumption, 21,* 57–77.

Branco, M. C., & Rodrigues, L. L. (2006). Corporate social responsibility and resource based perspectives. *J. Bus. Ethics, 69,* 111–132.

Braun, M., Latham, S., & Cannatelli, B. (2019). Strategy and business models: Why winning companies need both. *Journal of Business Strategy, 40*(5), 39–45.

Breitenfellner, A., Pointner, W., & Schuberth, H. (2019). The potential contribution of central banks to green finance. *Vierteljahrshefte zur Wirtschaftsforschung, 88,* 55–72.

Brisman, A. (2011). Stockholm conference, 1972. In D. K. Chatterjee (Eds.), *Encyclopedia of global justice.* Springer. https://doi.org/10.1007/978-1-4020-9160-5_655.

Brones, F., & de Carvalho, M. M. (2015). From 50 to 1: Integrating literature toward a systemic ecodesign model. *Journal of Cleaner Production, 96,* 44–57.

Brown, N., & Deegan, C. (1998). The public disclosure of environmental performance information—A dual test of media agenda setting theory and legitimacy theory. *Accounting and Business Research, 29*(1), 21–41.

Bruns, V., & Fletcher, M. (2008). Banks'risk assessment of Swedish SMEs. *Venture Capital, 10*(2), 171–194. https://doi.org/10.1080/13691060801946089.

Bux, H., Zhang, Z., & Ahmad, N. (2020). Promoting sustainability through corporate social responsibility implementation in the manufacturing industry: An empirical analysis of barriers using the ISMMICMAC approach. *Corporate Social Responsibility and Environmental Management.*

Buxel, H., Esenduran, G., & Griffin, S. (2015). Strategic sustainability: Creating business value with life cycle analysis. *Business Horizons, 58*(1), 109–122.

Campbell, S. (1996). Green cities, growing cities, just cities? Urban planning and the contradictions of sustainable development. *Journal of the American Planning Association, 62*(3), 296–312.

Carayannis, E. G., Sindakis, S., & Walter, C. (2015). Business model innovation as lever of organizational sustainability. *The Journal of Technology Transfer, 40*(1), 85–104.

Cardeal, G., Höse, K., Ribeiro, I., & Götze, U. (2020). Sustainable business models-canvas for sustainability, evaluation method, and their application to additive manufacturing in aircraft maintenance. *Sustainability, 12,* 9130. https://doi.org/10.3390/su12219130.

Carney, M. (2015, September). *Breaking the tragedy of the horizon—Climate change and financial stability.* Speech given at Lloyd's of London, 29. www.bankofengland.co.uk/publications/Pages/speeches/2015/844.aspx.

Carroll, A. B. (1979). A three-dimensional conceptual model of corporate social performance. *Academy of Management Review, 4,* 497–505.

Carroll, A. B. (1991). The pyramid of corporate social responsibility: Toward the moral management of organizational stakeholders. *Business Horizons, 34*(4), 39–48.

Carroll, A. B. (2016). Carroll's pyramid of CSR: Taking another look. *International Journal of Corporate Social Responsibility, 1*(1), 1–8.

Carroll, A. B. (1999). Corporate social responsibility evolution of a definitional construct. *Business & Society, 38*(3), 268–295.

Carter, C. R., & Rogers, D. S. (2008). A framework of sustainable supply chain management: Moving toward new theory. *International Journal of Physical Distribution & Logistics Management, 38*(5), 360–387. https://doi.org/10.1108/09600030810882816.

Casadesus-Masanell, R., & Ricart, J. E. (2010). From strategy to business models and onto tactics. *Long. Range Plan., 43,* 195–215.

Cassells, S., & Lewis, K. (2011). SMEs and environmental responsibility: Do actions reflect attitudes? *Corporate Social Responsibility and Environmental Management, 18,* 186–199.

Cavalcante, S., Kesting, P., & Ulhøi, J. (2011). Business model dynamics and innovation: (re)establishing the miss-ing linkages. *Management Decision, 49*, 1327–1342. https://doi.org/10.1108/00251741111163142.

Cetindamar, D., & Husay, K. (2007). Corporate social responsibility practices and environmentally responsible behaviour: The case of the United Nations Global Compact. *Journal of Business Ethics, 76*(2), 163–176.

CGC 2017: Coalition for Green Capital. (2007). National green banks in developing countries: Scaling up private finance to achieve paris. Climate goals coalition for green capital with support from the Hewlett foundation, July 2017.

Chambers, E., & Patrocínio, M. (2011). *Business models and value creation: A case study of The New York City economic development corporation*. Umeå School of Business.

Chan, H. L., Wei, X., Guo, S., & Leung, W. H. (2020). Corporate social responsibility (CSR) in fashion supply chains: A multi-methodological study. *Transportation Research Part E: Logistics and Transportation Review, 142*, 102063.

Chandler, D. (2019). *Strategic corporate social responsibility: Sustainable value creation*. Sage.

Chesbrough, H. W. (2007). Business model innovation: It's not just about technology anymore. *Strategy and Leadership, 35*, 12–17.

Chesbrough, H. W., & Rosenbloom, R. S. (2002). The role of the business model in capturing value from innovation: Evidence from Xerox Corporation's technology spinoff companies. *Industrial and Corporate Change, 11*, 533–534.

Chesbrough, H. (2007). Business model innovation: It's not just about technology anymore. *Strategic Leadership, 35*, 12–17.

Chiapello, E. (2018). The work of financialisation (Chapter 21). In I. Chambost, M. Lenglet, & Y. Tadjeddine (Eds.), *The making of finance: Perspectives from the social sciences*. Oxon, UK: Routledge.

Chladek, N. (2019, November 6). *Why you need sustainability in your business strategy, business insights*. https://online.hbs.edu/blog/post/business-sustainability-strategies.

Ciegis, R., Ramanauskiene, J., & Martinkus, B. (2009). The concept of sustainable development and its use for sustainability scenarios. *Inzinerine Ekonomika-Engineering Economics, 2*(62), 28–37.

Čihák, M., Demirgüç-Kunt, A., Feyen, E., Levine, R. (2012). Benchmarking financial systems around the world.The World Bank, Financial and Private Sector Development Vice Presidency & Development Economics Vice Presidency. Available online https://openknowledge.worldbank.org/bitstream/handle/10986/12031/wps6175.pdf.

Collevecchio. (2003). *Implementing the Collevecchio declaration on financial institutions and sustainability.*https://www.publiceye.ch/de/news/detail/implementing-the-collevecchio-declaration-on-financial-institutions-and-sustainability. Accessed 2 Nov 2020.

Collevecchio. (2020). Collevecchio Dec*laration, The role and responsibility of financial institutions.* BankTrack. http://www.banktrack.org/download/collevechio_declaration/030401_collevecchio_declaration_with_signatories.pdf. Accessed 2 Nov 2020.

Communication from the Commission to the European Parliament, the European Council, the Council, the European Central Bank, the European Economic and Social Committee and the Committee of the Regions. Action Plan: Financing Sustainable Growth. European Commission, Brussels, 8.3.2018 COM(2018) 97 final.

Cornée, S., Kalmi, P., & Szafarz, A. (2019). *The business model of social banks.* Kyklos. 73. 10.1111/kykl.12221. https://www.researchgate.net/publication/338096801_The_Business_Model_of_Social_Banks. Accessed 16 Nov 2020.

Cornée, S., Kalmi, P., & Szafarz, A. (2016). Selectivity and transparency in social banking: Evidence from Europe. *Journal of Economic Issues, 50*(2), 494–502.

Cronin, J. J., Brady, M. K., & Hult, G. T. M. (2000). Assessing the effects of quality, value and customer satisfaction on consumer behavioral intentions in service environment. *Journal of Retailing, 76*(2), 193–218.

Cull, R., Haber S., & Imai M. (2011). *Related lending and banking development* (Policy Research Working Paper, No. 5570). The World Bank Development Research Group, Finance and Private Sector Development Team.

D'Andrea A., Ferri F., & Grifoni P. (2014). The E-commerce business model implementation. In J. Wang (Ed.), *Encyclopedia of business analytics and optimization* (pp. 2509–2520). IGI Global, Hershey.

Dahlsrud, A. (2008). How corporate social responsibility is defined. *Corporate Social Responsibility and Environmental Management, 15,* 1–13.

Dahlsrud, A. (2008). How corporate social responsibility is defined: An analysis of 37 definitions. *Corporate Social Responsibility and Environmental Management, 15*(1), 1–13.

Dawe, N. K., & Ryan, K. L. (2003). The faulty thee-legged-stool-model of sustainable development. *Conversation Biology, 24,* 1458–1460.

DiMaggio, P. J., & Powell, W. W. (1983). The iron cage revisited: Institutional isomorphism and collective rationality in organizational fields. *American Sociological Review,* 147–160.

Dembek, K., York, J., & Singh, P. J. (2018). Creating value for multiple stakeholders: Sustainable business models at the base of the pyramid. *Journal of Cleaner Production, 196,* 1600–1612.

Dempsey, N., Bramley, G., Power, S., & Brown, C. (2011). The social dimension of sustainable development: Defining urban social sustainability. *Sustainable Development, 19*(5), 289–300.

Demil, B., & Lecocq, X. (2010). Business model evolution: In search of dynamic consistency. *Long Range Planning, 43*, 227–246. https://doi.org/10.1016/j.lrp.2010.02.004.

den Ouden, E. (2011). *Innovation design: Creating value for people, organizations and society*. Springer Science & Business Media.

den Ouden, E. (2012). *Innovation design: Creating value for people, Organizations and Society*. London: Springer.

Dikau, S., & Ryan-Collins, J. (2017). *Green central banking in emerging market an developing country economies*. New Economics Foundation.

Dixit, A. (2009). Governance institutions and economic activity. *American economic review, 99*(1), 5–24.

Dohrmann, S., Raith, M., & Siebold, N. (2015). Monetizing social value creation—A business model approach. *Entrepreneurship Research Journal, 5*(2), 127–154.

Doleski, O. D. (2015). *Integrated business model: Applying the St. Gallen Management concept to business models*. Springer Gabler.

Doualle, B., Medini, K., Boucher, X., & Laforest, V. (2015). Investigating sustainability assessment methods of product-service systems. *Procedia Cirp, 30*, 161–166.

Dunbar, R. L. M., & Starbuck, W. H. (2006). Learning to design organizations and learning from designing them. *Organization Science, 17*, 171–178.

Dyck, B., & Silvestre, B. S. (2018). Enhancing socio-ecological value creation through sustainable innovation 2.0: Moving away from maximizing financial value capture. *Journal of Cleaner Production, 171*, 1593–1604.

Dyllick, T., & Muff, K. (2016). Clarifying the meaning of sustainable business: Introducing a typology from business-as-usual to true business sustainability. *Organization & Environment, 29*(2), 156–174.

Dyllick, T., & Rost, Z. (2017). Towards true product sustainability. *Journal of Cleaner Production, 162*, 346–360.

Dziawgo, L. (2010). *Zielony rynek finansowy*. PWE Warszawa: Ekologiczna ewolucja rynku finansowego.

EC Action Plan on financing sustainable growth, UE Commision. (2018). https://www.unpri.org/download?ac=5173.

EC Europe. (2019). *Paris climate conference (COP21)*. https://ec.europa.eu/clima/policies/international/negotiations/paris_en. Accessed 31 Dec 2020.

Edwards, A. R. (2005). *The sustainability revolution: Portrait of a paradigm shift*. New Society Pub.

Elkington, J. (1994). Towards the sustainable corporation: Win win strategies for sustainable development. *California Management Review, 36*(2), 90–100.

Elkington, J. (1994). Towards the sustainable corporation: Win-win-win business strategies for sustainable development. *California management review, 36*(2), 90–100.

Elkington, J. Cannibals with Forks: The Triple Bottom Line of 21st Century Business; New Society Publishers: Gabriola, BC, Canada, 1998; ISBN 978-0-86571-392-5.

Engert, S., Rauter, R., & Baumgartner, R. J. (2016). Exploring the integration of corporate sustainability into strategic management: A literature review. *Journal of Cleaner Production, 112,* 2833–2850.

Environmental, social and governance integration for banks: A guide to starting implementation. Ed. J. Stampe, World Wide Fund For Nature (Formerly World Wildlife Fund), Gland, Switzerland, 2014, p. 12.

Epstein, G. A. (2009). *Post-war experiences with developmental central banks: The good, the bad and the hopeful* (G-24 Discussion Paper No. 54). Geneva: UNCTAD.

ES Europe 2019, EC Europa. (2019). *EU Sustainable finance taxonomy—Technical report.* https://ec.europa.eu/info/sites/info/files/business_economy_euro/banking_and_finance/documents/190618-sustainable-finance-teg-report-taxonomy_en.pdf). Accessed 2 Nov 2020.

Escrig-Olmedo, E., Ángeles; Fernández-Izquierdo, M.; Ferrero-Ferrero, I., Rivera-Lirio, J. M., & Muñoz-Torres M. J. (2019). Rating the raters: Evaluating how ESG rating agencies integrate sustainability principles. *Sustainability, 11,* 915.

Evans, S., Rana, P., & Short, S. W. (2014). Final set of tools & methods that enable analysis of future oriented, novel, sustainable, value adding business-models and valuenetworks. Deliverable D2. 6, Project 262931, SustainValue: Sustainable value creation in manufacturing networks.

Evans, S., Vladimirova, D., Holgado, M., Van Fossen, K., Yang, M., Silva, E. A., et al. (2017). Business model innovation for sustainability: Towards a unified perspective for creation of sustainable business models. *Business Strategy Environment, 26*(5), 597–608.

Evans, S., Vladimirova, D., Holgado, M., Van Fossen, K., Yang, M., Silva, E. A., & Barlow, C. Y. (2017). Business model innovation for sustainability: Towards a unified perspective for creation of sustainable business models. *Business Strategy and the Environment*, *26*(5), 597–608.

Ferenhof, H. A., Vignochi, L., Selig, P. M., Lezana, Á. G. R., & Campos, L. M. S. (2014). Environmental management systems in small and medium-sized enterprises: An analysis and systematic review. *Journal of Cleaner Production, 74,* 44–53.

Fielt, E. (2013). Conceptualising business models: Definitions, frameworks and classifications. *Journal of Business Models, 1*(1), 85–105.

Filipiak, B., & Dylewski, M.. (2020). Sustainable financial reporting in the context of ensuring sustainability of financial systems In M. Ziolo (Ed.), *Finance and sustainable development* (pp. 235–267). Routledge.

Filipiak, B. Z., & Dylewski, M. (2020). Sustainable Financial Reporting in the Context of Ensuring Sustainability of Financial Systems. In M. Ziolo (Ed.), *Finance and Sustainable Development* (pp. 235–267). York: Routledge, London-New.

Firm Finance, *Journal of Business*, 68, 351–381.

Forstater, M., Halle, M., & Zade S. (2016). *Green finance for developing countries: Needs, concerns and innovations*. United Nations Environment Programme.

Foss, N. J., & Saebi, T. (2015). Business models and business model innovation: Bringing organization into the field. In N. J. Foss & T. Saebi (Eds.), *Business model innovation: The organisational dimension* (pp. 1–23). Oxford: Oxford University Press.

Foss, N. J., & Saebi, T. (2017). Fifteen years of research on business model innovation: How far have we come, and where should we go? *Journal of Management, 43*(1), 200–227.

França, C. L., Broman, G., Robert, K. H., Basile, G., & Trygg, L. (2017). An approach to business model innovation and design for strategic sustainable development. *Journal of Cleaner Production, 140,* 155–166.

Freudenreich, B., Lüdeke-Freund, F., & Schaltegger, S. (2019). A stakeholder theory perspective on business models: Value creation for sustainability. *Journal of Business Ethics, 40*(1), 98.

Fullwiler, S. T., (2015). *Sustainable finance: Building a more general theory of finance* (Working Paper No. 106).

Gao, P., & Li, J. (2020). Understanding sustainable business model: A framework and a case study of the bike-sharing industry. *Journal of Cleaner Production, 267*, 122229.

Garavan, T. N., Heraty, N., Rock, A., & Dalton, E. (2010). Conceptualizing the behavioral barriers to CSR and CS in organizations: A typology of HRD interventions. *Advances in Developing Human Resources, 12*(5), 587–613.

García-Muiña, F. E., Medina-Salgado, M. S., Ferrari, A. M., & Cucchi, M. (2020). Sustainability transition in Industry 4.0 and Smart Manufacturing with the Triple-Layered Business Model Canvas. *Sustainability, 12*(6), 2364.

García-Sánchez, I. M., & García-Sánchez, A. (2020). Corporate social responsibility during COVID-19 pandemic. *Journal of Open Innovation: Technology, Market, and Complexity, 6*(4), 126.

García-Sánchez, I. M., & García-Sánche, A. (2020). Corporate social responsibility during COVID-19 pandemic. *Journal of Open Innovation: Technology, Market, and Complexity, 6,* 126. https://doi.org/10.3390/joitmc6040126.

Gassmann, O., Frankenberger, K., & Csik, M. (2014). *The business model navigator: 55 models that will revolutionise your business.* Pearson UK 2010.

Gassmann, O., Frankenberger, K., & Csik, M. (2010). *The business model navigator*. Pearson..

Geissdoerfer, M., Vladimirova, D., & Evans S. (2018). Sustainable business model innovation: A review. *Journal of Cleaner Production, 198,* 401416.

Geissdoerfer, M., Bocken, N. M., & Hultink, E. J. (2016). Design thinking to enhance the sustainable business modelling process—A workshop based on a value mapping process. *Journal of Cleaner Production, 135,* 1218–1232.

Geissdoerfer, M., Morioka, S. N., de Carvalho, M. M., & Evans, S. (2018). Business models and supply chains for the circular economy. *Journal of Cleaner Production, 190,* 712–721.

Geissdoerfer, M., Savaget, P., & Evans, S. (2017). The Cambridge business model innovation process. *Procedia Manufacturing, 8,* 262–269.

Geissdoerfer, M., Savaget, P., Bocken, N. M. P., & Hultink, E. J. (2017). The circular economy a new sustainability paradigm? *Journal of Cleaner Production, 143,* 757–768.

Gerster, R. (2012). *Sustainable finance: Achievements, challenges, outlook.* Striking a Balance Ahead of Rio+20in 2012. Available online http://www.gersterconsulting.ch/docs/sustainable_finance_final_11.02.10.pdf.

Giesen, E., Berman, S., Bell, R., & Blitz, A. (2007). Paths to success, three ways to innovate your business model. *Journal of Strategy and Leadership, 35*(6), 27–33.

Gilding, P., Hogarth, M., & Don Reed, D. (2002). Single bottom lline sustainability: How a values centred approach to corporate sustainability can pay off for shareholders and society (Sydney, Australia: Ecos Corporation), 18–19.

Girotra, K., & Netessine, S. (2013). OM forum—business model innovation for sustainability. *Manufacturing & Service Operations Management, 15*(4), 537–544.

Global4. (2020). *Business model canvas – Global4net*. https://global4net.com/ecommerce/business-model-canvas-co-to-jest-i-jak-go-stworzyc/.

González-Benito J., & González-Benito, Ó. (2005). A study of the motivations for the environmental trans-formation of companies. *Industrial Marketing Management*, 34 (7), 462–475.

Goyal, P., & Kumar, D. (2017). *Modeling the CSR barriers in manufacturing industries*. Benchmarking: An International Journal.

Gray, S., & Talbot, N. (2007). *Developing financial markets*. London, UK: Bank of England.

Griffith, A., & Bhutto, K. (2008). Improving environmental performance through integrated management systems (IMS) in the UK. *Management of Environmental Quality: An International Journal, 19*(5), 565–578.

Grosvold, J., Hoejmose, S. U., & Roehrich, J. K. (2014). Squaring the circle: Management, measurement and performance of sustainability in supply chains. *Supply Chain Management: An International Journal, 19*(3), 292–305.

Guo, H., Zhao, J., & Tang, J. (2013). The role of top managers' human and social capital in business model innovation. *Chinese Management Studies, 7*(3), 447–469.

Guo, Y. (2017). Implementing relationship banking strategies and techniques and improving customer value. *Finance and Market, 2*(2), 1–13.

Gupta, H. (2017). Integration of quality and innovation practices for global sustainability: An empirical study of Indian SMEs. *Global Business Review, 18*(1), 210–225.

Hahn, R., & Kühnen, M. (2013). Determinants of sustainability reporting: a review of results, trends, theory, and opportunities in an expanding field of research. *Journal of Cleaner Production, 59,* 5–21.

Hall, J., & Vredenburg, H. (2003). The challenges of innovating for sustainable development. *MIT Sloan Management Review, 45*(1), 61–68.

Hall, J. K., Daneke, G. A., & Lenox, M. J. (2010). Sustainable development and entrepreneurship: Past contributions and future directions. *Journal of Business Venturing, 25*(5), 439–448.

Hart, S. (1997, January–February). Beyond greening: Strategies for a sustainable world. *Harvard Business Review*, 66–76.

Hart, S. L., & Milstein, M. B. (2003). Creating sustainable value. *Academy of Management Executive, 17*(2), 56–69.

Hart, S. L., Milstein, M. B., & Caggiano, J. (2003). Creating sustainable value. *Academy* of *Management Executive, 17*(2), 56–67.

Figge, F., & Hahn, T. (2004). Sustainable value addeddmeasuring corporate contributions to sustainability beyond eco-efficiency. *Ecological Economics, 48*(2), 173–187.

Henriques, I., & Sadorsky, P. (1999). The relationship between environmental commitment and manage-rial perceptions of stakeholder importance. *Academy of Management Journal, 42*(1), 87–99.

Hess, D., & Warren, D. E. (2008). The meaning and meaningfulness of corporate social initiatives. *Business and Society Review, 113*(2), 163–197.

Hockerts, K. (2015). A cognitive perspective on the business case for corporate sustainability. *Business Strategy and the Environment, 24,* 102–122.

Holgado Granados, M. (2014). *Sustainable value creation in manufacturing through maintenance services*, PhD thesis. University of Milan.

Holtström, J., Bjellerup, C., & Eriksson, J. (2019). Business model development for sustainable apparel consumption: The case of Houdini Sportswear. *Journal of Strategy and Management, 12*(4), 481–504. https://doi.org/10.1108/JSMA-01-2019-0015.

Hörisch, J. (2015). The role of sustainable entrepreneurship in sustainability transitions: A conceptual synthesis against the background of the multi-level perspective. *Administrative Sciences, 5,* 286–300.

https://ec.europa.eu/info/business-economy-euro/banking-and-finance/sustainable-finance/eu-taxonomy-sustainable-activities_en.

https://www.bankinghub.eu/banking/research-markets/esg-investments-current-trends.

https://www.forbes.com/sites/timothyjmcclimon/2020/01/02/5-corporate-social-responsibility-trends-to-follow-in-2020/#4474e6987e69.

https://www.ifm.eng.cam.ac.uk/research/industrial-sustainability/sustainable-business-models/tools/cambridge-value-mapping-tool/.

Huang, Y. (2011). Is economic volatility detrimental to global sustainability? World Bank Economic Review, *26,* 128–146.

Hughes, R. C., & Troy, M. (2017). Motivations of one small business community to implement environmental sustainability as a business practice. *Journal of Environmental Sustainability,* 5(1), Article 2, 1–8, Available at http://scholarworks.rit.edu/jes/vol5/iss1/2.

Hurtado Jaramillo, C. H., Arimany Serrat, N., Meijide Vidal, D., & Ferràs Hernández, X. (2018). *Sustainability values for business: A perspective of value alignment in a supplier–client relationship for case aqualogy.*

Janicka, M. (2016). Financial markets and the challenges of sustainable growth, comparative economic research. *Central and Eastern Europe, 19*(2), 27–41.

Jesson, J., Matheson, L., & Lacey, F. M. (2011). *Doing your literature review: Traditional and systematic techniques.* Sage.

Jeucken, M. (2001). *Sustainable finance and banking: The financial sector and the future of the planet.* Earthscan Publications.

Jha, N., & Bhome, S. (2013). A study of green banking trends in India. *International Monthly Referred Journal of Research in Management and Technology, 2,* 127–132.

Johnson, M. W. (2010). *Seizing the white space: Business model innovation for growth and renewal.* Boston: Harvard Business Press.

Johnson, M. W., Christensen, C. M., & Kagermann, H. (2008). Reinventing your business model. *Harvard Business Review, 86*(12), 50–59.

Johnson, D., Straker K., Cara, W., & Bucolo, S. (2013). Designing innovative business models: Five emerging meta-models. In Cai, Jun, Lockwood, Thomas, Wang, Chensheng, Tong, Gabriel Y., & Liu, Jikun (Eds.), Proceedings 2013 IEEE Tsinghua International Design Management Symposium : Design-Driven Business Innovation, Institute of Electrical and Electronics Engineers, Inc., Shenzhen, China, 70–77.

Johnson, M. P., & Schaltegger, S. (2016). Two decades of sustainability management tools for SMEs: How far havewe come? *Journal of Small Business Management, 54,* 481–505.

Jones, D., & Gregor, S. (2006). *The formulation of an information systems design theory for e-Learning*. Presented at the First International Conference on Design Science Research in Information Systems and Technology, Claremont, CA, 356–373.

Joyce, A., & Paquin, L. R. (2016). The triple layered business model canvas: A tool to design more sustainable business models. *Journal of Cleaner Production, 135,* 1474–1486.

Joyce, A., & Paquin, R. L. (2016). The triple layered business model canvas: A tool to design more sustainable business models. *Journal of cleaner production, 135,* 1474–1486.

Joyce, A., & Paquin, R. L. (2016). The triple layered business model canvas: A tool to design more sustainable business models. *Journal of cleaner production, 135,* 1474–1486.

Joyce, A., Paquin, R., & Pigneur, Y. (2015). The triple layered business model canvas: A tool to design more sustainable business models. In *ARTEM Organizational Creativity International Conference,* 26–27 March 2015, Nancy, France.

Kaplan S., & Winby, S. (2007). Organizational models for innovation. Available online http://www.vps.ns.ac.rs/Materijal/mat936.pdf. Accessed 23 Sept 2019.

Kenig-Witkowska, M. (2017). The concept of sustainable development in the European Union Policy and Law. *Journal of Comparative Urban Law and Policy, 1*(1), 64–80. Study Space IX Cape Town, South Africa Article 6.

Khan, S. Z., Yang, Q., & Waheed, A. (2019). Investment in intangible resources and capabilities spurs sustainable competitive advantage and firm performance. *Corporate Social Responsibility and Environmental Management, 26,* 285–295.

Kiernan, P. (2014, September 23). *Risky business, The economist intelligence unit*. www.economistinsights.com/energy/opinion/risky-business.

Kiron, D., Kruschwitz, N., Haanaes, K., & von Streng Velken, I. (2012). Sustainability nears a tipping point. *MIT Sloan Management Review, 53*(2), 69–74.

Kleindorfer, P. R., Singhal, K., & Van Wassenhove, L. N. (2005). Sustainable operations management. *Production and Operations Management, 14*(4), 482–492.

Klimontowicz, M., & Pyka, A. (2016). Banks' business models – The challenges of XXI century. *Studia i Prace WNEiZ US, 44*(2), 153–165. https://doi.org/10.18276/sip.2016.44/2-12.

Knyphausen-Aufsess, D. Z., & Meinhardt Y. (2002). Revisiting Strategy: ein Ansatz zur Systematisierung von Gesch€aftsmodellen. In T. Bieger, N. Bickhoff, R. Caspers, D. Zu Knyphausen-Aufseß, & K. Reding (Eds.), *Zukünftige Gesch€aftsmodelle*. Springer.

Kolk, A., & Mauser, A. (2002). The evolution of environmental management: From stage models to performance evaluation. *Business strategy and the environment, 11*(1), 14–31.

KPMG International. (2011, December). *Evolving banking regulation. A long journey ahead – the outlook for 2012.* Report.

Kramer, M. R., & Porter, M. (2011). *Creating shared value* (Vol. 17). FSG.

Kreibiehl S., & Patel S. (2014, March). *Delivering the green economy through financial policy.* Frankfurt School of Finance & Management—UNEP Collaborating Centre for Climate & Sustainable Energy Finance, 2014.

Krick, T. et al. (2005). *From words to action the stakeholder engagement manual volume 2: The practitioner's handbook on stakeholder engagement.* Available at http://www.accountability.org/images/content/2/0/208.pdf.

Księżak, P. (2016). The benefits from CSR for a company and society. *Journal of Corporate Responsibility and Leadership, 3*(4), 53–65.

Kudłak, R., Szőcs, I., Krumay, B., & Martinuzzi, A. (2018). The future of CSR-selected findings from a Europe-wide Delphi study. *Journal of Cleaner Production, 183,* 282–291.

Kumar, V., Rahman, Z., & Kazmi, A. A. (2016). Assessing the influence of stakeholders on sustainability marketing strategy of Indian companies. *SAGE Open, 6*(3), 1–19.

Kurucz, E. C., Colbert, B. A., Luedeke-Freund, F., Upward, A., & Willard, B. (2017). Relational leadership for strategic sustainability: Practices and capabilities to advance the design and assessment of sustainable business models. *Journal of Cleaner Production, 140,* 189–204.

Laasch, O. (2019). An actor-network perspective on business models: How 'Being Responsible' led to incremental but pervasive change. *Long Range Planning, 52*(3), 406–426.

Lagoarde-Segot, T. (2020). Financing the sustainable development goals. *Sustainability, 12,* 2775.

Lankoski, L., & Smith, N. C. (2018). Alternative objective functions for firms. *Organization and Environment, 31*(3), 242–262.

Laudal, T. (2011). Drivers and barriers of CSR and the size and internationalization of firms. *Social Responsibility Journal, 7*(2), 234–256.

Laukkanen, M., & Tura, N. (2020). The potential of sharing economy business models for sustainable value creation. *Journal of Cleaner Production, 253,* 120004.

Leal Filho, W., & Brandli, L. (2016). Engaging stakeholders for sustainable development. In W. Leal Filho & L. Brandli (Eds.), *Engaging stakeholders in education for sustainable development at University Level, World Sustainability Series* (pp. 335–342).

Leaton, J., Ranger, N., Ward, B., Sussams, L., & Brown, M. (2013). *Unburnable carbon 2013: Wasted capital and stranded assets.* London: Carbon Tracker

&The Grantham Research Institute, London School of Economics and Political Science.

Lee, J., Suh, T., Roy, D., & Baucus, M. (2019). Emerging technology and business model innovation: The case of artificial intelligence. *Journal of Open Innovation: Technology, Market, and Complexity, 5*, 44.

Lenssen, G., Painter, M., Ionescu-Somers, A., Pickard, S., Bocken, N., Short, S., … & Evans, S. (2013). *A value mapping tool for sustainable business modelling*. Corporate Governance.

Lindic, J., & Marques (2011). Value proposition as a catalyst for a customer focused innovation. *Management Decision, 49*(10), 1694–1708. https://doi.org/10.1108/00251741111183834.

Lindgardt, Z., Reeves, M., Stalk, G., & Deimler, M. (2009). *Business model innovation: When the game gets tough change the game*. The Boston Consulting Group. http://www.bcg.com/documents/file36456.pdf.

Liston-Heyes. C., & Ceton, G. (2007). Corporate social performance and politics, do liberals Do More? *The Journal of Corporate Citizenship, 25*, Spring, 95–108.

Louche, C., Idowu, S. O., & Filho, W. L. (2010). *From risk management to value creation*. Greenleaf Publishing.

Lozano, R. (2008). Envisioning sustainability three-dimensionally. *Journal of Cleaner Production, 16*(17), 1838–1846.

Lozano, R. (2012). Towards better embedding sustainability into companies' systems: An analysis of voluntary corporate initiatives. *Journal of Cleaner Production, 25*, 14–26.

Lozano, R. (2014). Creativity and organizational learning as means to foster sustainability. *Sustainable Development, 22*(3), 205–216.

Lüdeke-Freund, F. (2010). Towards a conceptual framework of business models for sustainability. In *Proceedings of ERSCP-EMSUConference 2010—Knowledge collaboration and learning for sustainable innovation. European Roundtable on Sustainable Consumption and Production Society*. Ed by Wever, R., J. Quist, A. Tukker, J. Woudstra, F. Boons, and N. Beute.

Lüdeke-Freund, F., Carroux, S., Joyce, A., Massa, L., & Breuer, H. (2018). The sustainable business model pattern taxonomy—45 patterns to support sustainability-oriented business model innovation. *Sustainable Production and Consumption, 15*, 145–162.

Lüdeke-Freund, F., Gold, S., & Bocken, N. M. (2019). A review and typology of circular economy business model patterns. *Journal of Industrial Ecology, 23*(1), 36–61.

Luttropp, C., & Lagerstedt, J. (2006). EcoDesign and The Ten Golden Rules: Generic advice for merging environmental aspects into product development. *Journal of Cleaner Production, 14*(15–16), 1396–1408.

Macklem, T. (2011, June 6). *Mitigating systemic risk and the role of central banks*. Remarks to "Conférence de Montréal", Montreal, Quebec. www.bis.org/review/r110610c.pdf.

Macquarie, R. (2018). *A Green Bank of England*. Partners for a New Economy: Positive Money.

Mahajan, R., & Bose, M. (2018). Business sustainability: Exploring the meaning and significance. *IMI Konnect, 7*(2), 8–13.

Marcel, J. (2001). *Sustainable finance and banking: The financial sector and the future of the planet*. Earthscan.

Markides, X. (1999). *Adapted from All the Right Moves*. Boston: Harvard Business School Press.

Matten, D., & Moon, J. (2008). "Implicit" and "explicit" CSR: A conceptual framework for a comparative understanding of corporate social responsibility. *Academy of Management Review, 33,* 404–424.

Mazzucato, M., & Penna, C. C. R. (2015). *Beyond market failures: The market creating and shaping roles of state investment banks*. Working paper. New York: Levy Economic Institute of Bard College.

McWilliams, A., & Siegel, D. (2000). Corporate social responsibility and financial performance: correlation or misspecification? *Strategic Management Journal, 21*(5), 603–609.

Mil, G. (2006). The financial performance of a socially responsible investment over time and a possible link with corporate social responsibility. *Journal of Business Ethics, 63*(2), 131–148.

Mishra, D. K. (2013). Green strategies: Response of Indian Banks to climate change. *The Ecoscan, 3,* 345–348.

Moon, J. (2007). The contribution of corporate social responsibility to sustainable development. *Sustainable Development, 15*(5), 296–306.

Morioka, S. N., & de Carvalho, M. M. (2016). A systematic literature review towards a conceptual framework for integrating sustainability performance into business. *Journal of Cleaner Production, 136,* 134–146.

Morioka, S. N., Bolis, I., & de Carvalho, M. M. (2018). From an ideal dream towards reality analysis: Proposing Sustainable Value Exchange Matrix (SVEM) from systematic literature review on sustainable business models and face validation. *Journal of Cleaner Production, 178,* 76–88.

Morioka, S. N., Bolis, I., Evans, S., & Carvalho, M. M. (2017). Transforming sustainability challenges into competitive advantage: Multiple case studies kaleidoscope converging into sustainable business models. *Journal of Cleaner Production, 167,* 723–738.

Morioka, S. N., Bolis, I., & Carvalho, M. M. D. (2018). From an ideal dream towards reality analysis: Proposing Sustainable Value Exchange Matrix (SVEM) from systematic literature review on sustainable business models and face validation. *Journal of Cleaner Production, 178,* 76–88.

Morris, M., Schindehutte, M., & Allen, J. (2005). The entrepreneur's business model: Toward a unified perspective. *Journal of Business Research, 58*, 726–735.

Mullins, J., & Komisar, R. (2009). *Getting to plan B: Breaking through to a better business model*. Boston, MA: Harvard Business School Press.

Mustafa, R., & Werthner, H. (2011). Business models and business strategy—Phenomenon of explicitnes. *International Journal of Global Business & Competitiveness, 6*(1), 15–30.

NGFS. (2019). First comprehensive report: A call for action. Climate Change as a Source of Financial Risk. Network for Greening the Financial System. Paris.

Nosratabadi, S., Pinter, G., Mosavi, A., & Semperger, S.(2020). Sustainable banking; evaluation of the European business models. *Sustainability*, *12*, 2314.

Nosratabadi, S., Pinter, G., Mosavi, A., & Semperger, S. (2020). Sustainable banking; evaluation of the European business models. *Sustainability, 12*, 2314.

Olawumi, T. O., & Chan, D. W. A. (2018). scientometric review of global research on sustainability and sustainable development. *Journal of Cleaner Production, 183*, 231–250.

Ongena, S., & Smith, D. C. (2000). Bank relationships: A survey. In P. Haker & S. A. Zenios (Eds.), *The performance of financial institutions* (pp. 221–258). London: Cambrige University Press.

ONZ. (2015). *Transforming our World: The 2030 agenda for sustainable development*. United Nations 2015, A/RES/70/1; https://sdgs.un.org/sites/default/files/publications/21252030%20Agenda%20for%20Sustainable%20Development%20web.pdf.

Osborn, D., Amy Cutter, A., & Ullah, F. (2015). *Universal sustainable development goals understanding the transformational challenge for developed countries*. Stakeholder Forum. https://sustainabledevelopment.un.org/content/documents/1684SF_-_SDG_Universality_Report_-_May_2015.pdf.

Osterwalder, A. (2004). *The business model ontology—A proposition in a design science approach*. Dissertation 173, University of Lausanne, Switzerland.

Osterwalder, A., Ondrus J., & Pigneur, Y. (2005, May). *Skype's disruptive potential in the telecom market: A systematic comparison of business models*. HEC LAUSANNE Working paper.

Osterwalder A., & Pigneur, Y. (2010). *Business model generation: A handbook for visionaries, game changers, and challengers*. John Wiley & Sons.

Osterwalder, A., & Pigneur, Y. (2002). *An eBusiness model ontology for modeling eBusiness*. In Bled 2002, Proceedings, AIS Electronic Library, 75–91.

Osterwalder, A., & Pigneur, Y. (2009). *Business model generation*. Amsterdam: Selfpublished.

Osterwalder, A., & Pigneur, Y. (2010). *Business model generation—A Handbook 131 for visionaries, game changers and challengers*. Wiley.

Osterwalder, A., & Pigneur, Y. (2012). Designing business models and similar strategic objects: The contribution of IS. *Journal of the Association for Information Systems, 14*(5) , Article 3. https://doi.org/10.17705/1jais.00333. Available at https://aisel.aisnet.org/jais/vol14/iss5/3.

Osterwalder, A., & Pigneur, Y. (2013). Designing business models and similar strategic objects: The contribution of IS. *Journal of AIS, 14*(5), 237.

Osterwalder, A., Pigneur, Y., Oliveira, M. A. Y., & Ferreira, J. J. P. (2011). Business model generation: A handbook for visionaries, game changers and challengers. *African Journal of Business Management, 5*(7), 22–30.

Osterwalder, A., Pigneur, Y., & Tucci, C. L. (2005). Clarifying business models: Origins, present, and future of the concept. *Communications of the Association for Information Systems, 16*(1), 1.

Osterwalder, A., Pigneur, Y., & Clark, T. (2010). *Business model generation*. Wiley.

Palo, T., & Tähtinen, J. (2014). A network perspective on business models for emerging technology-based services. *Journal of Business & Industrial Marketing, 26*(5), 377–388.

Park, H., & Kim, J. D. (2020). Transition towards green banking: Role of financial regulators and financial institutions. *Asian Journal of Sustainability and Social Responsibility, 5*(5), 1–25. https://doi.org/10.1186/s41180-020-00034-3.

Patala, S., Jalkala, A., Keränen, J., Väisänen, S., Tuominen, V., & Soukka, R. (2016). Sustainable value propositions: Framework and implications for technology suppliers. *Industrial Marketing Management, 59*, 144–156.

Pearce, D. W., & Atkinson, G. D. (1993). Capital theory and the measurement of sustainabledevelopment: an indicator of "weak" sustainability. *Ecological Economics, 8*(2), 103–108.

Peat, R. O. S. S. (2003). Values drive value. *University of Auckland Business Review, 5*(2), 1–11.

Pellicano, M., Vesci, M., Troisi, O., Cosimato, S., & Tuccillo, C. (2014). *Stakeholder engagement for sustainable development in banking industry: An Italian case study*. In 13th EBES conference—Istanbul, Turkey, pp 1–17. Available at http://www.researchgate.net/publication/275887421_Stakeholder_engagement_for_sustainable_development_in_banking_industry._An_Italian_case_study.

Peltola, T., Aarikka-Stenroos, L., Viana, E., & Mcakinen, S. (2016). Value capture in business ecosystems for municipal solid waste management: Comparison between two local environments. *Journal of Cleaner Production, 137*, 1270–1279.

Perspektywy rozwoju zrównoważonego finansowania – implikacje dla sektora przedsiębiorstw finansowych, Deloitte 2019.
Petersen, M. A., & Rajan, R. G. (1994). The benefits of lending relationships: Evidence fromsmall business data. *The Journal of Finance, 49*(1), 3–37. https://doi.org/10.2307/2329133.
Porter, M., & Kramer, M. (2011). Creating shared value. *Harvard Business Review, 89*(1/2), 62–77.
Porter, M. E., & Kramer, M. R. Creating shared value. In G. G. Lenssen, N. C. Smith (Eds.), *Managing sustainable business: An executive education case and textbook* (pp. 323–346). Springer. ISBN 978-94-024-1144-7.
PRI. (2020). *The Un principles for responsible investment and the OECD guidelines for multinational enterprises: Complementarities and distinctive contributions.*http://www.oecd.org/investment/mne/38783873.pdf. Accessed 2 Nov 2020.
Qastharin, A. R. (2015). *Business model canvas for social enterprise*. The 7th Indonesia International Conference on Innovation, Entrepreneurship, and Small Business (IICIES 2015). https://www.researchgate.net/publication/323393037_Business_Model_Canvas_for_Social_Enterprise.
Rai, R., Kharel, S., Devkota, N., & Paudel, U. R. (2019). Customers perception on green banking practices: A desk. *Journal of Economic Issues, 10,* 82–95.
Ramdani, B., Binsaif, A., & Boukrami, E. (2019). Business model innovation: A review and research agenda. *New England Journal of Entrepreneurship, 2*(2), 79–108.
Randles, S., & Laasch, O. (2016). Theorising the normative business model. *Organization & Environment, 29*(1), 53–73.
Raport 2020: Zielone finanse - nowe spojrzenie na finansowanie. (2020). Klub Odpowiedzialnych Finansów przy Europejskim Kongresie finansowym, Sopot.
Rauter, R., Jonker, J., & Baumgartner, R. J. (2017). Going one's own way: Drivers in developing business models for sustainability. *Journal of Cleaner Production, 140,* 144–154.
Renard, Y. J. S. (2015). A review of corporate sustainability reporting tools (SRTs). *Journal of Environmental Management, 164,* 180–195.
Reynolds, S. J. (2019). Overcoming the psychological barriers of CSR. *Organizational Dynamics*, No. 1/100745. ISSN 0090-2616. https://doi.org/10.1016/j.orgdyn.2019.100745.
Richardson, B. J. (2005). Equator principles, The voluntary approach to environmentally sustainable finance. *The European Environmental Law Review, 14,* 280.
Richardson, B. J. (2009). Keeping ethical investment ethical: Regulatory issues for investing for sustainability. *Journal of Business Ethics, 87*(4), 555–572.
Richardson, J. E. (2005). *The business model: An integrative framework for strategy execution*. Available at SSRN 932998.

Roome N., & Louche, C. (2015). Journeying toward business models for Sustainabil-ity: A conceptual model found inside the black box of organi-sational trans-formation. *Organization and environment*. Sage.

Roome, N., & Louche, C. (2016). Journeying toward business models for sustainability: A conceptual model found inside the black box of organisational transformation. *Organization & Environment, 29*(1), 11–35.

Rosca, E., Arnold, M., & Bendul, J. C. (2017). Business models for sustainable innovation an empirical analysis of frugal products and services. *Journal of Cleaner Production, 162,* 133–145.

Russo, M. V., & Fouts, P. A. (1997). A resource-based perspective on corporate environmental performance and profitability. *The Academy of Management Journal, 40*(3), 534–559.

Sabatier, V., Mangematin, V., & Rousselle, T. (2010). From recipe to dinner: Business model portfolios in the European biopharmaceutical industry. *Long Range Planning, 43*(2–3), 431–447.

Sadovska, V., Ekelund Axelson, L., & Mark-Herbert, C. (2020). Reviewing value creation in agriculture—A conceptual analysis and a new framework. *Sustainability, 12*(12), 5021.

Saebi, T., & Foss, N. J. (2014). Business models for open innovation: Matching heterogenous open innovation strategies with business model dimensions. *European Management Journal, 3*(3), 201–213.

Sakis, K., Pinney, C. H., & Serafeim, G. (2016). ESG integration in investment management: Myths and realities. *Journal of Applied Corporate Finance, 28*(2) (Spring 2016), 10–16.

San-Jose, L., Retolaza, J. L., & Gutierrez, J. (2011). Are ethical banks different? A comparative analysis using the radical affinity index. *Journal of Business Ethics, 100*(1), 151–173.

San-Jose, L., Retolaza, J. L., & Gutierrez, J. (2011). Are ethical banks different? A comparative analysis using the radical affinity index". *Journal of Business Ethics, 100,* 151–173.

Santos,J., Spector B., & van der Heyden, L. (2009). *Toward a theory of business model innovation with incumbent firms*. Working paper no. 2009/16/EFE/ST/TOM. Fontainebleau: INSEAD, 5–13.

Sarkar, S., & Searcy, C. (2016). Zeitgeist or chameleon? A quantitative analysis of CSR definitions. *Journal of Cleaner Production, 135,* 1423–1435.

Savitz, A. W., & Weber, K. (2007). The sustainability sweet spot. *Environmental Quality Management, 17*(2), 17–28.

Schaltegger, S., Lüdeke-Freund, F., & Hansen, E. (2012). Business cases for sustainability: The role of business model innovation for corporate sustainability. *International Journal of Innovation & Sustainable Development, 6,* 95–119.

Schaltegger, S., Hansen, E. G., & Lüdeke-Freund, F. (2016). *Business models for sustainability: Origins, present research, and future avenues.*

Schaltegger, S., Hansen, E., & Lüdeke-Freund, F. (2016). Business models for sustainability: Origins, present research, and future avenues. *Organization & Environment, 29,* 3–10. https://doi.org/10.1177/1086026615599806.

Schaltegger, S., Lüdeke-Freund, F., & Hansen, E. G. (2011). *Business cases for sustainability and the role of business model innovation: developing a conceptual framework.* Centre for Sustainability Management (CSM), Leuphana Universität Lüneburg.

Schoenmaker, D., & Tilburg, R. V. (2016). What role for financial supervisors in addressing environmental risks? *Comparative Economic Studies, 58,* 317–334.

Scholtens, B., & van't Klooster, S. (2019). Sustainability and bank risk. *Palgrave Commun, 5,* 105. https://doi.org/10.1057/s41599-019-0315-9.

Schoneveld, G. C. (2020). Sustainable business models for inclusive growth: Towards a conceptual foundation of inclusive business. *Journal of Cleaner Production, 277,* 124062.

Schutten, W. (2016), *Sustainable banking business models.* Research Report. Master Business Administration, University of Twente.

Seuring, S., & Müller, M. (2008). From a literature review to a conceptual framework for sustainable supply chain management. *Journal of Cleaner Production, 16*(15), 1699–1710.

Shihadeh F., Gamage, S. K. N., & Hannoon, A. M. T.(2019). The causal relationship between SME sustainability and banks' risk. *Economic Research-Ekonomska Istraživanja, 32*(1), 2743–2760. https://doi.org/10.1080/1331677X.2019.1655465.

Short, S. W., Bocken, N. M. P., Rana, P., & Evans, S. (2012). Business model innovation for embedding sustainability: A practice-based approach introducing business model archetypes. In *10th Global Conference on Sustainable Manufacturing.*

Shrivastava, P., & Statler, M. (Eds.). (2012). *Learning from the global financial crisis: creatively, reliably, and sustainably.* Stanford University Press

Siebenhüner, B., & Arnold, M. (2007). Organizational learning to manage sustainable development. *Business Strategy and the Environment, 16*(5), 339–353.

Šimberová I., & Kita P.(2020). New business models based on multiple value creation for the customer: A case study in the chemical industry. *Sustainability, 12*(9), 3932. https://doi.org/10.3390/su12093932.

Slávik, Š., & Bednár, R. (2014). Analysis of business models. *Journal of Competitiveness, 6*(4):19–40.

Slywotzky, X. (1995). *Value migration.* Boston: Harvard Business School Press.

Sohn, L. B. (1973). The Stockholm declaration on the human environment. *The Harvard International Law Journal, 14*(3), 423–514.

Sonnemann, G., & Margni, M. (2015). *Life cycle management* (p. 353). Springer Nature.

Spieth, P., Schneckenberg, D., & Ricart, J. E. (2014). Business model innovation—State of the art and future challenges for the field. *R&d Management, 44*(3), 237–247.

Stankeviciene, J., & Nikonorova, M. (2014). Sustainable value creation in commercial banks during financial crisis, contemporary issues in business, management and education 2013. *Procedia—Social and Behavioral Sciences, 110,* 1197–1208.

Steen, M. (2013). Co-design as a process of joint inquiry and imagination. *Des Issues, 29*(2), 16–28.

Stern, N. (2016). *Climate change and central banks*. Bank for International Settlements.

Strategy for Smart, Sustainable and Inclusive Growth. European Commission, Communication of the European Commission 2020, Brussels, 3.3.2010, KOM(2010) 2020 final, www.ec.europa.eu/ eu2020/ pdf/1_PL_ACT_part1_v1.pdf.

Stubbs, W., & Cocklin, C. (2008). Conceptualizing a "sustainability business model". *Organization & Environment, 21*(2), 103–127.

Su, H., Rabbani, M. R., Ahmad, N., Sial, M. S., Cheng, G., Zia-Ud-Din, M., et al. (2020). CSR, co-creation and green consumer loyalty: Are green banking initiatives important? A moderated mediation approach from an emerging economy. *Sustainability, 12,* 10688. https://doi.org/10.3390/su122410688.

Sustainable Development. (2017). https://academic.oup.com/ejil/article/23/2/377/487236/Sustainable-Development-in-International-Law. Accessed 2 Nov 2020.

Sustainable finance taxonomy—Regulation (EU) 2020/852.

Tan, A. R., Matzen, D., McAloone, T. C., & Evans, S. (2010). Strategies for designing and developing services for manufacturing firms. *CIRP Journal of Manufacturing Science and Technology, 3*(2), 90–97.

Tanwar, R. (2013). Porter's generic competitive strategies. *IOSR Journal of Business and Management, 15*(1), 11–17.

Tara, K., Singn, S., & Kumar, R. (2015). Green banking for environmental management: A paradigm shift. *Current World Environment, 10*(3), 1029–1038.

Taxonomy report: technical annex, Updated methodology & Updated Technical Screening Criteria, march 2020.

Teece, D. (2010). Business models, business strategy and innovation. *Long Range Planning, 43,* 172–194.

TEG final report on the EU taxonomy, Taxonomy: Final report of the Technical Expert, Group on Sustainable Finance, European Commission, March 2020

Tikkanen, H., Lamberg, J.-A., Parvinen, P., & Kallunki, J.-H. (2005). Managerial cognition, action and the business model of the firm. *Management Decision, 43*, 789–809. https://doi.org/10.1108/00251740510603565.

The Equator Principles. (2013). *The Equator Principles June 2013, A financial industry benchmark for determining, assessing and managing environmental and social risk in projects*. www.equator-principles.com. Accessed 2 Nov 2020.

Tokarski, A., Tokarski, M., & Wójcicki, J.(2017). The possibility of using the business model canvas in the establishment of an operator's business plan. *Torun Business Review, 16*(4), 17–31.

Trimi, S., & Berbegal-Mirabent, J. (2012). Business mode linnovation in entrepreneurship. *The International Entrepreneurship and Management Journal, 8*(4), 449–465.

Tura, N., Keränen, J., & Patala, S. (2019). The darker side of sustainability: Tensions from sustainable business practices in business networks. *Industrial Marketing Management*, *77*, 221–231.

UE. (2018). *Communication from the commission to the European Parliament, the European Council, the Council, the European Central Bank, the European Economic and Social Committee and the Committee of the Regions Action Plan: Financing Sustainable Growth*. Document 52018DC0097; COM/2018/097 final. https://eur-lex.europa.eu/legal-content/PL/TXT/?uri=COM:2018:0097:FIN.

UNEP. (2016). *Definitions and concepts: Background note*. UNEP.

UNEP Inquiry. (2016). Greening the banking system—Taking stock of G20 green banking market practice. Geneva: UN Environment Inquiry into the Design of a Sustainable Financial System.

UNEPFI. (2016). *Connecting financial system and sustainable development: Market leadership paper* (p. 3). https://www.unepfi.org/wordpress/wp-content/uploads/2016/11/MKT-LEADERSHIP-REPORT-AW-WEB.pdf. Accessed 2 Nov 2020.

United Nations. (2019). *About the sustainable development goals*. Available at https://www.un.org/sustainabledevelopment/sustainable-development-goals/. Accessed 12 2020.

Upward, A., & Jones, P. (2016). An ontology for strongly sustainable business models: Defining an enterprise framework compatible with natural and social science. *Organization & Environment, 29*(1), 97–123.

Uschold, M., & King M. (1995). *Towards a methodology for building ontologies*. Citeseer.

Values based banking. Bringing the voice of the citizen into finance. UNEP Inquiry, IWP 15/03, 2015, pp. 6. http://unepinquiry.org/wp-content/uploads/2015/04/Values_Based_Banking.pdf. Accessed 5 Jan 2020.

Voelpel, S. C., Leibold, M., Tekie, E. B., & von Krogh, G. (2005). Escaping the red queen effect in competitive strategy: Sense-testing business models.

European Management Journal, 23, 37–49. https://doi.org/10.1016/j.emj.2004.12.008.

Van Wassenhove, L. N. (2019). Sustainable innovation: Pushing the boundaries of traditional operations management. *Production and Operations Management, 28*(12), 2930–2945.

Vandaele, N. J., & Decouttere, C. J. (2013). Sustainable R&D portfolio assessment. *Decision Support Systems, 54*(4), 1521–1532.

Villena, V. H. (2019). The missing link? The strategic role of procurement in building sustainable supply networks. *Production and Operations Management, 28*(5), 1149–1172.

Visser, W. (2010). The age of responsibility: CSR 2.0 and the new DNA of business. *Journal of Business Systems, Governance and Ethics, 5*(3), 7.

Visser, W. (2011). *The age of responsibility: CSR 2.0 and the new DNA of business*. John Wiley & Sons.

Visser, W. (2016). The future of CSR: Towards transformative CSR, or CSR 2.0. In *Research handbook on corporate social responsibility in context*. Edward Elgar Publishing.

Volz, U. (2017). *On the role of central banks in enhancing green finance, United Nations environment programme*. Inquiry Working Paper 17/01.

Watson, D. (2005). *Business models*. Petersfield: Harriman House Ltd.

Weber O. (2016). *The impact of green banking: Guidelines on the sustainability performance of banks*. The Chinese Case Policy Brief, 79, June 2016, 1–7.

Werner, R. A. (2003). *Princes of the Yen: Japan's central bankers and the transformation of the economy*. Eastgate Books.

What Is ISO 26000—Guidance on Social Responsibility? https://asq.org/quality-resources/iso-26000.

Windsor, Duane. (2006). Corporate social responsibility: Three key approaches. *Journal of Management Studies, 43*(1), 93–114.

Wirtz, B. W., Pistoia, A., Ullrich, S., & Göttel, V. (2016). Business Models: Origin, development and future research perspectives. *Long Range Planning, 49*(1), 36–54.

Witjes, S., & Lozano, R. (2016). Towards a more Circular Economy: Proposing a framework linking sustainable public procurement and sustainable business models. *Resources, Conservation and Recycling, 112*, 37–44.

Wonglimpiyarat, J. (2015). Challenges of SMEs innovation and entrepreneurial financing. *World Journal of Entrepreneurship, Management and Sustainable Development, 11*(4), 295–311. https://doi.org/10.1108/WJEMSD-04-2015-0019.

Wood, D. J. (1991). Corporate social performance revisited. *Academy of Management Review, 16*, 691–718.

World Commission on Environment and Development. (1987). *Our common future* (p. 27). Oxford: Oxford University Press. ISBN 019282080X.

Wu, H.-C., & Cheng, C.-C. (2018). An empirical analysis of green experiential loyalty: A case study. *Journal of International Food and Agribusiness Marketing, 31,* 69–105.

Wu, X., & Sivalogathasan, V. (2013). *Value business model for future: A complex organization to a simple organization.* 5th International Conference on "Role of Innovation in Business", IFIM Business School in Bangalore, India. 16th–17th May 2013, https://www.researchgate.net/publication/265844112_Value_Business_Model_for_Future_A_Complex_Organization_to_a_Simple_Organization.

www.1:https://www.un.org/sustainabledevelopment/sustainable-development-goals.

Yadlapalli, A., Rahman, S., & Gunasekaran, A. (2020). Corporate social responsibility definitions in supply chain research: An ontological analysis. *Journal of Cleaner Production, 277,*.

Yang, M., Evans, S., Vladimirova, D., & Rana, P. (2017). Value uncaptured perspective for sustainable business model innovation. *Journal of Cleaner Production, 140,* 1794–1804.

Yang, M., Evans, S., Vladimirova, D., & Rana, P. (2017). Value uncaptured perspective for sustainable business model innovation. *Journal of Cleaner Production, 140,* 794–804.

Yang, M., Vladimirova, D., & Evans, S. (2017). Creating and capturing value through sustainability: The sustainable value analysis tool a new tool helps companies discover opportunities to create and capture value through sustainability. *Research-Technology Management, 60*(3), 30–39.

Yang, M., Vladimirova, D., & Evans, S. (2017). Creating and capturing value through sustainability: The sustainable value analysis tool a new tool helps companies discover opportunities to create and capture value through sustainability. *Research-Technology Management, 60*(3), 30–39.

Yang, M., Vladimirova, D., Rana, P., & Evans, S. (2014). Sustainable value analysis tool for value creation. *Asian Journal of Management Science and Applications, 1*(4), 312–332.

Yang, M., Evans, S., Vladimirova, D., & Rana, P. (2017). Value uncaptured perspective for sustainable business model innovation. *Journal of Cleaner Production, 140,* 1794–1804.

Yip, A. W. H., & Bocken, N. M. P. (2018). Sustainable business model archetypes for the banking industry. *Journal of Cleaner Production, 174,* 150–169.

Zaman, G. (2012). Challenges and requirements for sustainable development of Romania'S agriculture based on the input-output analysis. *Romanian Journal of Economics, 35*(2)(44), 5–15.

Zhang, D., Morse, S., & Kambhampati, U. (2017). *Sustainable development and corporate social responsibility.* Routledge.

Zhang, Q., Oo, B. L., & Lim, B. T. H. (2019). Drivers, motivations, and barriers to the implementation of corporate social responsibility practices by construction enterprises: A review. *Journal of Cleaner Production, 210*, 563–584.

Zhelyazkova, V., & Kitanov, Y. (2015). Green banking-definition, scope and proposed business model, ecology & safety. *Journal of International Scientific Publications, 9*. https://www.researchgate.net/publication/330202134_GREEN_BANKING-DEFINITION_SCOPE_AND_PROPOSED_BUSINESS_MODEL/citation/download. Accessed 16 Nov 2020.

Zijp, M. C., Heijungs, R., Van der Voet, E., Van de Meent, D., Huijbregts, M. A., Hollander, A., et al. (2015). An identification key for selecting methods for sustainability assessments. *Sustainability, 7*(3), 2490–2512.

Zimmermann, S. (2019). Same same but different: How and why banks approach sustainability. *Sustainability, 11*, 2267. https://doi.org/10.3390/su11082267.

Ziolo, M., Filipiak, B. Z., Bąk, I., Cheba, K., Tîrca, D. M., & Novo-Corti, I. (2019). Finance, sustainability and negative externalities. An overview of the European context. *Sustainability, 11*(15), 4249. https://doi.org/10.3390/su11154249.

Ziolo, M., Filipiak, B. Z., Bąk, I., & Cheba, K. (2019). How to design more sustainable financial systems: The roles of environmental, social, and governance factors in the decision-making process. *Sustainability, 11*(20), 5604. https://doi.org/10.3390/su11205604.

Ziolo, M., Bąk, I., Cheba, K., & Spoz, A. (2020). Sustainable business models of enterprises-actual and declared activities for ensuring corporate sustainability. *Procedia Computer Science, 176*, 1497–1506.

Ziolo, M., Filipiak, B.Z.; Bąk, I., & Cheba, K. (2019). How to design more sustainable financial systems: The roles of environmental, social, and governance factors in the decision-making process. *Sustainability 2019, 11*, 5604.

Zott, C., & Amit, R. (2007). Business model design and the performance of entrepreneurial firms. *Organization Science, 18*, 181–199.

Zott, C., & Amit, R. (2008). The fit between product market strategy and business model: Implications for fi rm performance. *Strategic Management Journal, 29*(1), 1–26.

Zott, C., & Amit, R. (2010). Business model design: An activity system perspective. *Long Range Planning, Business Models, 43*(2–3), 216–226.

Zott, C., Amit, R., & Massa, L. (2011). The business model: Recent developments and future research. *Journal of Management, 37*, 1019–1042.

Index

© The Editor(s) (if applicable) and The Author(s), under exclusive license to Springer Nature Switzerland AG 2021

M. Ziolo et al., *Sustainability in Bank and Corporate Business Models*, Palgrave Studies in Impact Finance,
https://doi.org/10.1007/978-3-030-72098-8

The manufacturer's authorised representative in the EU is Springer Nature Customer Service Centre GmbH, Europaplatz 3, 69115 Heidelberg, Germany. If you have any concerns regarding our products, please contact ProductSafety@springernature.com

Printed and bound by CPI Group (UK) Ltd, Croydon, CR0 4YY
30/09/2026
02225701-0009